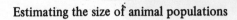
Estimating the size of animal populations

Estimating the size of animal populations

J Gordon Blower, Laurence M Cook
Department of Zoology, University of Manchester

James A Bishop
Department of Genetics, University of Liverpool

George Allen & Unwin Limited
London. Boston, Sydney.

First published in 1981

GEORGE ALLEN & UNWIN LTD
40 Museum Street, London WC1A 1LU

© J G Blower, L M Cook and J A Bishop 1981

This book was designed in collaboration with the Institute of Advanced Studies, Manchester Polytechnic.

British Library Cataloguing in Publication Data

Blower, J Gordon
    Estimating the size of animal populations.
    1. Animal populations – Statistical methods
    I. Title   II. Cook, Laurence   III. Bishop, James A
    591.5'2'4   QH352   79-41638

    ISBN 0-04-591017-0
    ISBN 0-04-591018-9 Pbk

Set in 10 on 11 point Press Roman at the Alden Press Ltd.
Oxford, London and Northampton,
and printed in Great Britain
by Biddles Ltd, Guildford, Surrey

Preface

Ecology and population genetics are concerned with the birth, death, numbers and movement of organisms. This book is a by-product of field courses in animal ecology, run over a number of years, on which we have tried to measure these variables. The first question posed when studying animals in a particular habitat is, 'How many are there?'. The question entails so many other interesting problems that the field studies have concentrated on it. Different habitats and types of animals require different methods, but even when trying to estimate the numbers of particular species in a single locality several possible methods exist. The investigator needs to know not simply what he should do but why there are alternative methods and what dictates their choice. We have tried to produce a handbook which outlines the procedures involved in calculation of population numbers, and also gives a concise account of the theory underlying each method. It should be useful as a basic working text for both undergraduate and graduate students. For those wishing to take the subject further, Southwood (1966, 1978) covers many practical aspects for insects, including trapping and marking as well as the methods of estimation, while the authoritative review of the theory of estimation is by Seber (1973). Our debt to the second source will be obvious.

We have stressed the underlying assumptions of the models so that the most appropriate method may be selected for the species and conditions under study. Rightly or wrongly, biologists working in the field tend to be rather sceptical about the value of standard errors of estimates, and prefer to look for independent confirmation of the values obtained. For this reason, as well as from limitation of space, we have said little about the error estimates. This aspect is covered fully by Seber.

The intention has been to make the treatment as free of context as possible. The examples used are terrestrial invertebrates and small mammals, but the methods may equally be applied to aquatic organisms. Daily samples have been taken, but different sample intervals may be used — a year is often practical for birds and molluscs. There is a large literature on tagging in fishery research. Some of it concerns estimation of other parameters such as movement or sustainable yield and some specialised techniques are required, but where population estimation is concerned the basic principles are those outlined here. Each type of organism presents its own problems of capture and marking. These are too many and varied to be included. Whatever the particular biological circumstances with which the reader has to deal, the book should indicate the logic of the statistical methods available and provide a guide to the steps in calculation.

We are grateful to the many students who have provoked discussion of population estimation and to Mr and Mrs A. R. Kelly of Woodchester Park Field Centre, Gloucestershire, where our courses have taken place. We have benefited from discussions with many of our colleagues; in particular we would like to thank Drs R. R. Baker, J. S. Bradley, P. D. Gabbutt, M. V. Hounsome, R. J. White, J. J. Whitelaw and D. W. Yalden. Drs G. J. Caughley, G. M. Jolly and J. J. Murray have kindly discussed some aspects of the subject matter with us.

Contents

Key to methods covered

1 Animals immobile or effectively so over the sampling period and pattern used	**Ch. 2** Area Samples
Animals mobile	2
2 Population effectively closed with no loss or gain	3
Population open	4
3 Animals killed and removed from the population or not considered after first capture and marking	**Ch. 5** Trapping and removal Change in ratio
Animals marked and recaptures scored	**Ch. 3** Lincoln Index **Ch. 5** Increase in fraction marked. Frequency of recapture
4 Mark release and recapture methods	**Chs 3** and **4** see Table 7.3 page 114, for details
Constant survival rate assumed over the sampling period	Fisher and Ford or in certain situations, Jackson
Variable survival	Jolly or in certain situations Manly and Parr

1 Introduction

Ecologists attempt to explain the numbers which occur in natural populations. It is often difficult or impossible to count all the individuals in a given place, so that many ingenious ways of estimating numbers from samples have been developed. We propose to give a commonsense introduction to these methods and to provide a guide to when and how they should be used.

The population

In this book a population is any collection of individuals of a stated species or of one age group or sex, or even of one stated physiological, genetic, or behavioural category within an age group of a species. The population inhabits a defined place during a given period of time. The place is usually further qualified as an area, volume or other spatial unit within a stated geographic area or habitat.

The sample

It is essential to understand what is involved in estimating the size of a population from samples. We can count all the barnacles on a given piece of rock without difficulty. The number of aphids on a sycamore leaf can be determined almost as easily, although some may drop off or fly away during the count. But determination of the numbers within the limited area involved is not the end point of the investigation. We might wish to compare the population of barnacles on sheltered and exposed shores. First we find a rock on a sheltered shore small enough to allow all individuals to be counted. Then we seek a similar rock at a similar tidal level on an exposed shore. But what sort of rock, with what aspect and with what surroundings? How shall we determine 'similar tidal level'? Clearly we need several rocks for each type of shore; they must be of known size or we must count within areas of known size on each rock. How do we choose the rocks? How do we specify the area on each rock? As scientists we are seeking generalisations concerning rocky and sheltered shores, not a comparison of two specific pieces of rock.

The aphid example reveals more clearly the limitations of specific counts. Why should we want to know the number of aphids on a leaf? This information is sought as a means to an end — the knowledge of the larger population of aphids on the tree or in a forest. The places where absolute counts can be made are merely *samples* from which generalisations are made about larger places. Two features prevent us from counting every animal in the larger place. First, the time, effort and concentration required are only available in exceptional cases. Secondly, a total count would almost certainly lead to disturbance and damage of the habitat and population we wish to study.

The two examples above are of sessile and of relatively immobile animals, both easy to count in limited areas. Most animals are either hidden and difficult to locate, or if visible, are active and highly mobile. In these circumstances complete counts of large samples may be impractical, and further problems of sampling are posed. Those encountered when sampling sessile or relatively immobile animals and when studying active, relatively mobile animals will now be examined in turn.

Sampling sessile or relatively immobile animals

A relatively immobile animal is one which is unlikely to enter or leave the sampling area during the time needed to take the sample. Animals are counted in a sample which is part of the larger area or habitat under investigation. The sample is an area (or volume or other specified spatial unit such as the sycamore leaf) within a larger area. We will call this larger area (the total area investigated) the sampling area; the area of the sample relative to this sampling area is the sample fraction. Ideally the sample would include a fraction of the total population exactly equal to the sample fraction. If we express the numbers of animals per unit sample space, we might then expect this density over the entire sampling area. Whether the samples represent the whole in this way is the main concern of sampling theory and practice. Difficulties arise because of the heterogeneity of most habitats and of the distribution of animals within them. As a result a sample is usually divided into parts, called sample units or quadrats, so as to take account of these sources of heterogeneity.

To secure a representative sample we need to have some knowledge of the pattern of heterogeneity. Sometimes the pattern is determined by visible factors such as plants and physical features, but at other times the factors generating the pattern may be hidden from view. Returning to the example of the aphids on the sycamore tree, where we can form a visual impression of heterogeneity, we would attempt to choose sample units (leaves, for example) from sparsely, moderately and densely populated areas of the crown. On the other hand, when estimating the density of lumbricid worms in a pasture we have no idea at the start whether their distribution is heterogeneous or not. Part of the object of the census might be to determine this pattern, but in any case we need the knowledge in order to spread the sample units equably over the sampling area. This apparent impasse is the central problem of sampling relatively immobile animals, and is discussed in Chapter 2.

Sampling relatively mobile animals

Animals such as winged insects, birds and small mammals present sampling problems by virtue of their mobility. An animal is 'relatively mobile' if it is able and likely to enter or to leave the sampling area during the sampling interval. A sample of such animals is a different entity from the static sample of the preceding section. Animals are captured within the entire sampling area and not just in a small subdivided portion of it. The sample clearly is part of the whole – but just what part is not immediately clear. In practice we might mark captured animals and release them, take a further sample at a later time and observe what fraction of the sample was marked, or we might persist with capturing until a decline in capture rate suggests that we have caught a substantial fraction. An idea of the sample fraction can be gained from the proportion of marks in a second sample or from the rate of decline of captures in successive samples. Knowing the number captured and the fraction this represents of the whole, we can estimate the size of the entire population.

Area samples and time samples

We shall refer to the static sample of relatively immobile animals as an area sample and that of relatively mobile animals as a time sample. In an area sample we make a complete count of part of the sampling area, whereas in a time sample we make a partial count of the complete sampling area. An estimate of population size may be made from a single area sample and is based on our knowledge of the sample fraction. If we wish to know whether the population is changing, successive samples may be taken and estimates derived from them. Populations of mobile animals, however, can only be estimated from two or more successive samples, since these are necessary to estimate the sampling fraction, which cannot be stated in terms of area. These basic principles are stated in Figure 1.1.

The bounds of the population estimated

The major problem encountered with samples of mobile animals concerns the area in which the animals are active. We know the area within which we sample or capture. Animals are present there at the time of the study, but we may not know whence they came or whither they are bound. The mark—recapture methods detailed in Chapter 3 enable us to calculate rates of immigration and emigration. These refer to the sampling area, into which and from which the animals move. However, the animals we handle may be active in a much wider area, the bounds of which are not known.

First catch your hare

The Victorian cookery expert, Mrs Beeton, is said to have begun one of her recipes in this way. A similar injunction must be given to the population ecologist. Location, capture and marking of animals pose many problems, but it is not part of the purpose of this book to discuss them. We should be aware, however, that the method of capture and extraction may well affect the interpretation which can be placed upon the estimates and that the theory must be related to the design of the sampling programme.

Figure 1.1
Area and time samples.

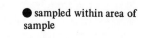
● sampled within area of sample

(a) Area sample: complete count of part of sampling area.

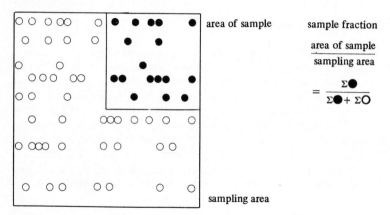

area of sample

sample fraction

$$\frac{\text{area of sample}}{\text{sampling area}}$$

$$= \frac{\Sigma \bullet}{\Sigma \bullet + \Sigma O}$$

sampling area

(b) Time sample: partial count of complete sampling area.

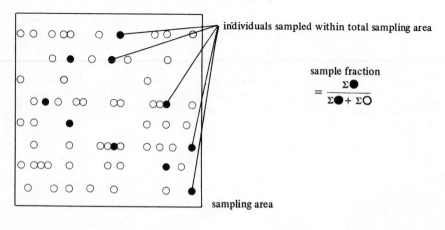

individuals sampled within total sampling area

sample fraction
$$= \frac{\Sigma \bullet}{\Sigma \bullet + \Sigma O}$$

sampling area

Figure 1.1 continued

The characteristics of area and time samples. Individual animals within the sampling area (large squares) are denoted by open and closed circles. Closed circles represent the animals captured or found in the sample. The area sample (a) is depicted as a smaller area equivalent to the rectangle marked *a* in Figure 2.1, but in practice this area is subdivided and the sample sub-units or quadrats are spread throughout the sampling area (see Ch. 2).

---

2 Area samples

Introduction

Population estimation by area samples is used for organisms which are relatively immobile during the course of the sampling procedure. The population must not change in size from the start to the finish of the sampling. The sample consists of an area, or sometimes a volume, of known size; all the organisms in this sample space are counted. The method is applicable to sessile animals such as barnacles or insect pupae and to mobile animals if they are removed as they are counted, or if the distance moved is sufficiently low as to make it impossible to count the same individual twice. Examples of mobile animals to which the method has been applied are caterpillars on herbage, small grasshoppers sampled in large sampling areas, wildebeest and other large migrating mammals and nesting song-birds holding territories. Animals such as protozoa or freshwater arthropod plankton have been studied by this method, even though they are sufficiently mobile to move out of the sampling unit (a volume of water), because all sampled individuals are removed from the population. Caughley (1977b) and Norton-Griffiths (1978) discuss the application of these methods to large mammals inhabiting a large area, where rapid sampling by aerial survey ensures that con-ditions are appropriate. Naturally, estimates based on area samples have been most extensively used in plant studies. The theory and practice are considered in detail by Grieg-Smith (1964, botanical examples), Elliott (1971, freshwater biology), Pielou (1977, theory), Seber (1973, statistical theory) and Southwood (1966, a general review). Cochran (1977) provides a most comprehensive cover-age of sampling theory.

Basic principles

The procedure is to measure the sampling area (**A**), measure the area of a portion of it, the sample (**a**), within which it is practical to count all the organisms, and then to count them to provide a number n (Fig. 2.1). The fraction of the total area examined is called the sample fraction (SF), and:

$$SF = \frac{\mathbf{a}}{\mathbf{A}} = \frac{\text{area of sample}}{\text{sampling area}}$$

We then make the assumption that the fraction of the population taken is equal to the fraction of the area examined, so that if P is the total population of the area:

$$\frac{n}{P} = \frac{a}{A} = SF \qquad (2.1)$$

so that:

$$\hat{P} = \frac{nA}{a} = \frac{n}{SF} \qquad (2.2)$$

The 'hat' on the P indicates that this is the estimate required. Compare this argument with that in Chapter 3, page 30.

The effect of pattern

The validity of this estimate of P depends on whether the area of the sample **a** is representative of the total area **A**. Figure 2.1 is drawn in such a way as to suggest intuitively that it is not, but how could the sampling procedure be improved? This depends on the pattern of distribution of individuals within **A**, which is usually unknown. As a result, estimation of P usually entails study of distributional pattern, and this may yield interesting results in its own right. In a similar manner estimation of mobile populations involves study of survival rates (cf. Ch. 3).

Figure 2.1
The basic sampling procedure.

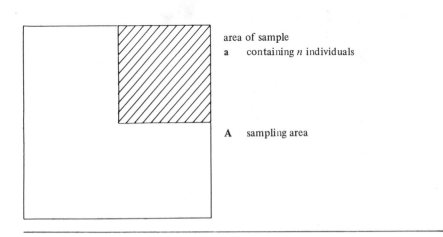

area of sample
a   containing n individuals

A   sampling area

For all living organisms the pattern of distribution in space falls somewhere on a continuum from regular through random to aggregated or patchy distribution. Examples of these distributions are shown in Figure 2.2. The validity of the sampling procedure shown in Figure 2.1 is clearly better for the regular distribution than for the other two.

The position of a species in this classification depends partly on the scale being used by the investigator. A species of beetle may be regularly distributed within clumps of vegetation that are randomly distributed in wetlands, which have a patchy distribution in a region of the country. The apparent pattern of distribution of the beetle will depend on whether the area of the sample unit is smaller than the vegetational clumps, or perhaps includes several of them. This consideration introduces a further level of complexity into the study of distribution. Without going into detail it is necessary to adopt a pattern of sampling which will ensure a representative sample.

The sampling procedure is usually based on a number of small sampling units, called quadrats when they are square. These must be of such a size that there is a good chance of each unit including one or more individuals. The problem then resolves itself into one of deciding where to place them. They may be distributed in a regular manner on a grid. This is often a satisfactory procedure, but it can result in bias if the organisms also have a regular pattern. It may also result in more intensive sampling of the edges of the area than of its centre. Random distribution of quadrats overcomes these problems. A random distribution may, however, include quite large empty areas, as in Figure 2.2b. This procedure can produce an unrepresentative sample if the distribution of the animals is affected by a fairly large-scale environmental patchiness. A good compromise is stratified random sampling. For this, the area is divided into a regular arrangement of sub-areas — strips or squares — and random sampling is carried out within each. The sub-areas should be chosen in such a way that the distribution of individuals within them is homogeneous. Sampling at random may then be undertaken in each. In practice, this implies that we have some information about the distribution, which may not be the case at the start of the sampling. Figure 2.3 shows a stratified random sampling procedure used for an exploratory study. The total area has been divided into 25 equally sized squares. When planning the sampling, the edges of each small square were marked off into 10 units and one quadrat was located within each square by finding random pairs of co-ordinates. These may be obtained by taking pairs of numbers between 0 and 9 from the tables of random numbers to be found in many statistical textbooks (some pocket calculators also have a random number facility). When examining the pattern of stratified random quadrats in Figure 2.3, the reader might justifiably ask how this differs from a regular arrangement. The difference is that every part of the sampling area is given an equal opportunity of being included within the area of the sample. Having established their layout, the quadrats are searched. The total number of animals found in each quadrat is then added together to give the number n, and the population of the area is estimated as in Equation 2.2. This method was used to investigate the number of earthworms in a meadow so that searching involved an extraction procedure as well.

Figure 2.2
Types of distribution of animals.

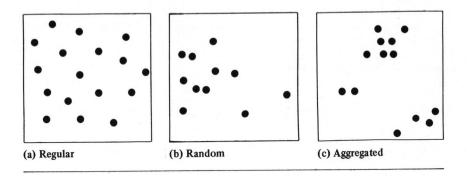

(a) Regular   (b) Random   (c) Aggregated

Figure 2.3
Stratified random distribution of quadrats.

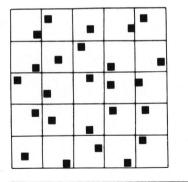

Accuracy of the estimate

Division of the area sampled into a number of quadrats allows the accuracy of the sampling techniques to be assessed. If the number of quadrats containing 0, 1, 2, 3 etc. individuals is plotted against the number they contain, we can often see at a glance whether a satisfactory sample has been obtained. If the histogram has a single mode and most quadrats contain 2, 3 or more animals, then a reasonable quadrat size and distribution have been adopted. However, if there is a bimodal distribution with a large number of quadrats containing no animals and a second peak at some high number of individuals per quadrat, then there is a risk that the sample taken is unrepresentative; the distribution of animals is aggregated and some large patches may have been missed or included by chance. This indicates that improvement might be effected by increasing the quadrat size or the number of quadrats, or both. Even when neither course is possible, an estimate of the accuracy of the estimate of P may be made.

If the distribution of individuals is random and the probability that they will fall within a quadrat is small, then the number of quadrats having $0, 1, 2, 3 \ldots$ x individuals should follow the Poisson distribution. The expected numbers for the Poisson distribution depend only on the mean number per quadrat, m, and may be calculated as follows:

Number of organisms/quadrat	0	1	2	3	x
Probability of getting this number	$\dfrac{1}{e^m}$	$\dfrac{m}{e^m}$	$\dfrac{m^2}{2e^m}$	$\dfrac{m^3}{6e^m}$	$\dfrac{m^x}{x!e^m}$

The expected number of quadrats is then N times the probability.

N total number of quadrats
n total number of organisms in the N quadrats
x number of organisms per quadrat
m mean number per quadrat $= n/N$
e base of natural logarithms $= 2.7183$

The observed numbers may be compared with the expected ones to give an indication of how the organisms are distributed. Over-representation in the zero category and in the high categories indicates clumping; over-representation in categories near the mean indicates regular distribution.

A measure of the spread of values about the mean is the variance of number per quadrat. This is found from:

$$V = \frac{1}{N-1}\left(\sum fx^2 - \frac{(\sum fx)^2}{N}\right)$$

V variance
f number of quadrats with a particular number x individuals.

Now it may be shown theoretically that for a Poisson distribution $V = m$. Consequently the fraction V/m is an index of aggregation. If it is less than 1 the distribution is regular, while values greater than 1 indicate clumping. The extent to which the distribution deviates from randomness may be tested using a χ^2 test calculated as follows:

$$\chi^2 = \frac{V}{m}(N-1)$$

This has N degrees of freedom. Notice that very low values of χ^2, as well as high ones, are of interest, since low values indicate significant deviation in the direction of regular dispersal.

If a set of data is examined and the χ^2 value is not significant, the investigator may accept the null hypothesis that $V = m$. In that case, the standard error of m is $\sqrt{(m/N)}$, so that the 95% confidence intervals for m are approximately $m - 2\sqrt{(m/N)}$ to $m + 2\sqrt{(m/N)}$. From (2.2) we have $P = n/SF$, but $n = mN$. To obtain the population estimate and its standard error (SE) both m and $\sqrt{(m/N)}$ are multiplied by N/SF. We therefore have:

$$\hat{P} = \frac{mN}{SF} \qquad\qquad (2.3)$$

$$SE_{\hat{P}} = \frac{\sqrt{mN}}{SF} \qquad\qquad (2.4)$$

By accepting the null hypothesis the investigator takes the risk of accepting as a random distribution one that deviates slightly, even though not significantly, from randomness. The alternative procedure would be to accept the calculated value of V as an estimate of the true variance of the distribution. In that case, the population estimate remains as (2.3) above, but its standard error becomes:

$$SE_{\hat{P}} = \frac{\sqrt{VN}}{SF} \qquad\qquad (2.5)$$

When the distribution is not random, alternative procedures are again open. Clumped distributions are very often fitted by the negative binomial distribution. For this both the mean and the variance are estimated. It is therefore possible to assume a negative binomial distribution, calculate the parameters and obtain the confidence intervals from them. This procedure will not be described here, but may be found, for example, in Southwood (1966), Debauche (1962) and Seber (1973) where indices of dispersion of various kinds are also discussed. Alternatively, the value of V may be calculated as described above and used to determine the standard error of P. If the negative binomial model is appropriate then this procedure overestimates the true variance. For our part, however, we prefer to use V when estimating a population, because it makes no assumptions about the distribution beyond those obtained directly from the variation in number of individuals between different quadrats.

The difference between the two methods is likely to be small. In a study of the larvae of the scarlet tiger moth, *Panaxia dominula*, 107 randomly distributed quadrats were searched. The mean number per quadrat was 4.98, with a variance V of 155.13. The distribution is very significantly different from a random distribution, and is clumped. A negative binomial distribution was fitted by a method given by Elliot (1971) and the difference between the observed and expected sequence was not significant. The parameter k of the distribution was estimated as 0.181. Using this value the variance is estimated as 142.00. The confidence intervals of the population estimate are reduced by 4% compared with the figure obtained using V. The investigator is fortunate when the sampling fraction is sufficiently high to make a difference of this scale meaningful.

Figure 2.4
Number of worms found in each quadrat in a trial sampling.

7	4	4	8	7
11	3	3	3	8
1	5	3	7	2
6	2	0	11	9
6	8	5	2	8

Example

An area of grass in a meadow was sampled to determine the number of worms of a particular type. The area was 15.8×15.8 m divided into 25 equally sized squares each of area $10 \, \text{m}^2$. A single $0.1 \, \text{m}^2$ quadrat was placed at random in each, as shown in Figure 2.3. The number of worms obtained is shown in Figure 2.4. These figures provide the following results:

$$SF = \frac{0.1}{10} = 0.01$$
$$N = 25$$
$$m = 5.32$$
$$V = 9.226$$

Figure 2.5 shows the histogram of observed number of quadrats containing different numbers of individuals, together with the Poisson distribution having the same mean. The observed distribution has a greater number of quadrats with low values and with high values than would be expected of a random distribution, indicating clumping of individuals. This is supported by the test of agreement, since:

$$\chi^2 = \frac{9.226 \times 24}{5.32} = 41.6$$

which with 25 degrees of freedom is significant at the 5% level. The null hypothesis that the distribution is a random one is therefore rejected. The standard error of the mean per quadrat is:

$$\sqrt{\frac{9.226}{25}} = 0.6075$$

Figure 2.5

Frequency distribution for the data in Figure 2.4 and the Poisson distribution expected if the distribution were random.

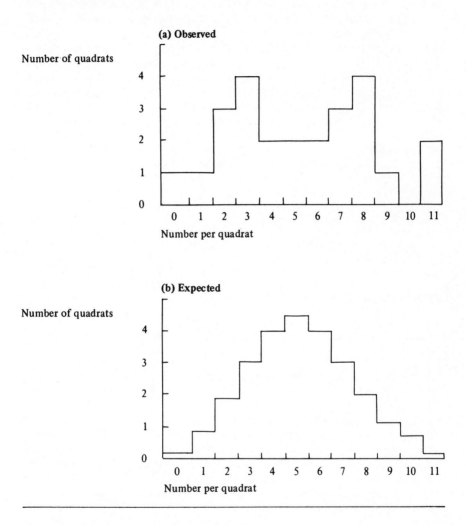

(a) Observed

Number of quadrats

(b) Expected

Number of quadrats

We therefore estimate the population as:

$$P = \frac{5.32 \times 25}{0.01} = 13\ 300 \text{ with standard error:}$$

$$SE_P = \frac{\sqrt{9.226 \times 25}}{0.01} = 1518.7$$

The 95% confidence intervals are approximately 10 263 to 16 337. These intervals are quite wide. The estimate would be improved by increasing the quadrat size, which might decrease the variance as well as increasing the sampling fraction, or simply by searching more quadrats. The survey did, however, show a significant difference between this area and another one in a nearby orchard.

Jolly (1969) and Caughley (1977b) discuss estimation of P when sampling units vary in size, as when data are collected during aerial transects across an irregularly shaped area. The method in the above example, if applied to such data, gives an unbiased but imprecise estimate of P. A more accurate estimate can be obtained if the relationship between number of animals seen and sample area is linear. Suppose there are N sample units, each of varying size z_i. The observer sees x_i animals in the i'th sample. Then:

A   total area
T   total number of units into which the area is divided
a   total area sampled $= \Sigma z$
n   total number of individuals seen $= \Sigma x$

As before, the population is estimated as:

$$\hat{P} = \frac{\mathbf{A}n}{\mathbf{a}}$$

This has the variance:

$$V = \frac{T^2}{N(N-1)}\left[\Sigma x^2 + \left(\frac{n}{\mathbf{a}}\right)^2 \Sigma z^2 - \frac{2n}{\mathbf{a}}\Sigma xz\right]$$

For this estimate and the one in Equation 2.3 N should not be much less than about 30. Caughley also discusses a method in which the probability of drawing a sampling unit is proportional to its size.

This basic procedure may be extended in several directions. For example Ashford, Read and Vickers (1970) discuss a model which permits the analysis of grasshopper and spittlebug populations. The basic data are obtained by area sampling on a series of occasions. A mathematical model is developed to represent the life history of the species and applied to the observed data. When employed for grasshoppers, their population sizes, rate of moult, duration of instars and death rate in each instar may be estimated. The process is quite complex and requires access to a computer. Discussion of such methods is beyond the scope of this book, but it is necessary to emphasise that area sampling is a powerful tool which can be made to yield a wide variety of information about natural animal populations. References will be found in the books referred to in the introduction to this chapter.

3 Time samples: basic principles of mark, release and recapture

Introduction

In the next three chapters we deal with the rationale and procedure for estimating numbers from time samples. Relatively mobile animals are those able and likely to enter or leave the sampling area during the sampling interval — animals which refuse to 'stand still' to be counted. The problem created by mobility is overcome by marking the animals as we count them and releasing the marked individuals back into the population. The population estimate is then obtained by observing the fraction of marked animals recaptured in one or more subsequent samples. These mark, release and recapture procedures will subsequently be referred to as MRR methods.

Animals in a time sample are gathered from the entire sampling area and not just from a set of prescribed sub-areas or quadrats. Because of this, the estimate of the sample fraction (the fraction of the entire population within the sampling area which we take) can only be derived from the proportion of marked animals or marks recaptured in a subsequent sample. A time sample therefore includes at least two separate capturing occasions.

If we were to attempt to take an area sample of relatively mobile animals, we would be in danger of counting the same animal more than once in a quadrat, or the same animal might be counted in several quadrats. To overcome this difficulty we could isolate captured animals in a keep net or cage until the count was complete, but this might lead to new individuals entering the area to fill the space provided. Animals might also restrict their movements to definite times of the day or night and different age groups might have different times of day for flying or running. Furthermore, the capture of an individual in a given area will not tell us whether the individual is a resident of the area, a regular visitor or a rare migrant. Marking the animals allows us to keep a tally of individuals that have been counted. By following the different proportions of recaptured marks in subsequent samples we can estimate the number habitually occupying or visiting the area, since data on rates of movement into and out of the area arise as natural by-products of the MRR exercise. The variables of number, survival and activity are all closely interrelated and the value of an MRR programme is that all these variables can be disentangled and separately estimated. Area samples of relatively immobile animals give us periodic estimates of density and age structure. By studying a series of such samples we can derive from them estimates of survival and movement. In a time sample, survival and activity are estimated at the same time as density; indeed they form part of the process of estimating density.

Mobile animals may be sampled directly during their inactive period. An obvious method of estimating the breeding population of a species of bird is to take an area sample of occupied nests. Many of the cryptozoa remain inactive by day under surface refuges, and an area sample can be taken of these. Difficulties stem from the fact that refuges tend to shelter aggregates of different sizes, but even if the distribution of refuges is reasonably homogeneous and scattered, a further problem posed by such animals is their low density in relation to a workable sample unit or quadrat. For example, we have student project estimates of the density of the common woodland carabid beetle *Nebria brevicollis* of about one per square metre. Now the litter within a square metre in which 0, 1 or 2 beetles may be taking refuge is a very large volume to search effectively. In fact, most animals able to displace themselves from a manageable quadrat of, say, a square metre will be found to be too sparsely distributed for area sampling; grasshoppers, frequently the subjects of class MRR exercises in Britain, usually occur here at densities between 0.1 and 1.0 per square metre. MRR methods are therefore appropriate for highly mobile animals which are sparsely represented in a quadrat of reasonable size. Marking allows us to cope with their mobility; in turn, their mobility removes the need for us to search large quadrats for small numbers of individuals.

The first analysis of MRR data is usually attributed to Lincoln (1930), although Le Cren (1965) suggests that it should be credited to Petersen (1889), while Laplace (1783) used the method to estimate the population of France in the eighteenth century from the register of births and the number of births in parishes of known size. The more complicated analysis of sequential MRR data was developed by Jackson (1933 *et seq.*), Fisher (see Dowdeswell, Fisher and Ford 1940; Fisher and Ford 1947) and Schnabel (1938). Jackson was estimating densities of tsetse flies with a view to the practical business of controlling the disease carried by the flies; Fisher and Ford were concerned with changes in gene frequency and the evolution of populations. Modifications for specific purposes were made by Leslie and Chitty (1951), Darroch (1961), Hammersley (1953) etc. Later, MRR theory was further developed by Jolly (1965), Seber (1965) and Manly and Parr (1968). These methods rely on multiple recaptures, unlike the earlier methods of Jackson and Fisher and Ford. A review of the mathematics of the methods is given by Seber (1973) and by Cormack (1969, 1973). Parr (1965) describes the application of several of the methods in a population study. Some authors refer to the basic calculation as the Petersen estimate, because it is neither an index nor was it originated by Lincoln whose name, however, appears to have stuck. Laplace used the same calculation, although he did not mark individuals, and no doubt he had other predecessors. We have therefore continued to call it the Lincoln Index; similarly, we refer to the other MRR methods by the name of a single author or pair of authors although in some cases others were also involved. The principal MRR methods will now be described within the context of simple double and triple catches for ease of understanding, and later (Ch. 4) in the detail of a longer series of samples.

The aim is to show how the underlying logic operates and to find a common-sense path through the mathematics.

The principle of the Lincoln Index

(a) Argument in terms of simple proportion

Animals are captured, marked and released back into the population. A second sample is taken after a period of time, say one day, and the number of marked animals in the sample (recaptures) is noted. In other words, we first mark an unknown fraction of the population and, on a second occasion, we take a sample of this fraction. Provided the effect of sampling error is negligible we can now write:

fraction of population captured, $=$ fraction of second sample which are
marked and released   recaptures of marked animals

$$\frac{n_i}{P_i} = \frac{m_{i+1}}{n_{i+1}} \tag{3.1}$$

where: P_i population size on day i
         n_i animals captured, marked and released on day i
         n_{i+1} number of animals in second sample taken day $i + 1$
         m_{i+1} number of recaptured marked animals in the second sample

By simple rearrangement of (3.1) we have:

$$P_i = \frac{n_i n_{i+1}}{m_{i+1}} \tag{3.2}$$

The conditions for the truth of identities (3.1) and (3.2) are:

(i)   Marked animals must evenly intermingle with the unmarked.
(ii)   Marked animals must behave in every respect like the unmarked; in particular, they must suffer no damage and must not be subject to greater risk of predation or capture than the unmarked.
(iii)   There must be no addition of individuals (by birth or immigration) to the population between occasions i and $i + 1$ since this will dilute the fraction of marked animals in the population.

Loss from the population will not alter the fraction n_i/P_i because marked animals will move out or die at the same rate as the unmarked, provided condition (ii) is fulfilled.

(b) Sampling intensity on the first day

If we seek animals in an area, the number we find and capture will evidently depend upon the time and effort we apply to the task and also on the efficiency with which we operate. Efficiency will depend on our ability to see the animals against different backgrounds, our knowledge of the kinds of microhabitat they occupy and our skill in countering their escape reactions. Since our efficiency is likely to be less than 100%, and time and effort will be limited, we shall catch only a fraction of the total population; this fraction is the sample fraction *SF* and is the result of a given sampling intensity. We can therefore write:

$$SF_i = \frac{n_i}{P_i} = \frac{m_{i+1}}{n_{i+1}} \tag{3.3}$$

provided there are no additions to the population between i and $i + 1$.

(c) Sampling fraction on the second occasion: argument in terms of the absolute number of marks available for recapture

The sampling fraction on the second day need not be the same as on the first for the truth of the identities (3.1) and (3.2). In fact, the sampling fraction on the second day is n_{i+1}/P_{i+1}. An estimate of this fraction can only be made if there is no loss from the population between days i and $i + 1$. The estimate of the sampling fraction on the second day is then:

$$\frac{\text{actual marks recaptured}}{\text{marks available to be recaptured}} = \frac{m_{i+1}}{n_i}$$

Thus we can write:

$$SF_{i+1} = \frac{n_{i+1}}{P_{i+1}} = \frac{m_{i+1}}{n_i} \tag{3.4}$$

and by rearrangement of the last two terms:

$$P_{i+1} = \frac{n_i n_{i+1}}{m_{i+1}} \tag{3.5}$$

The conditions for the truth of identities (3.4) and (3.5) are:

(i)
(ii) } as previously
(iii) There must be no loss (by death or emigration) of individuals between occasions i and $i + 1$. In particular, there must be no loss of marked individuals. The total marks in the population, n_i must remain the same until

$i + 1$ otherwise the right hand term in (3.4) would not be a true measure of SF_{i+1}. Addition of individuals to the population will not alter the truth of equation 3.4.

(d) Résumé of basic principles

The right-hand terms in Equations 3.2 and 3.5 are identical; we may refer to this quantity, $n_i n_{i+1}/m_{i+1}$, as the **Lincoln Index**. If the population is neither losing nor gaining, the index gives an estimate of either P_i or P_{i+1}; if there is loss and no gain, the index gives an estimate for P_i; if there is gain but no loss, the index gives an estimate for P_{i+1}.

Another way of expressing the difference between the Equations 3.2 and 3.5 is to stress that they use different estimators for the sampling fraction; Equation 3.2 uses an estimator for SF_i whereas Equation 3.5 uses an estimator for SF_{i+1}. For a constant sampling effort, loss from the population from day i to day $i + 1$ will have led to a proportionate reduction of both recaptured marks (m_{i+1}) and marks available for recapture (n_{i+1}) thus ensuring that m_{i+1}/n_{i+1} remains a true measure of SF_i. If the population is growing, the representation of recaptures in the sample (m_{i+1}) will be smaller but the number of marks available (n_i) will not have increased and the fraction m_{i+1}/n_i will be unchanged.

These basic principles are explained by a concrete example in Figure 3.1 but note that here we have kept the sampling fraction constant. Now we have two methods of estimating the size of a population, one for a population to which individuals are added and another for a population from which animals are subtracted; what more do we need? Unfortunately, we do not know whether our population is gaining, losing, or gaining and losing at the same time.

If loss and gain occur together, the proportion of marks and the absolute numbers of marks are changed and Equation 3.2 and Equation 3.5 will be invalid. However, while neither of the population estimates will be permissible, certain of the arguments leading up to them are still true, in particular the statements that only gain can alter the ratio m_{i+1}/n_{i+1} in (3.2) and that only loss can alter the absolute number of marks n_i in (3.5), are still true. If we had a method of detecting these changes we would be in a position to make an accurate estimate of the population. Any such method requires at least one more sampling session. We shall now describe the principles of the methods used, with reference to a triple capture scheme.

The triple catch

(a) The models

The principles of the simple Lincoln Index have been developed with reference to a double catch: animals first captured were marked and released and the number of recaptures in the second catch was noted. Now suppose the animals

Figure 3.1

Simple mark recapture with static, declining and growing populations.

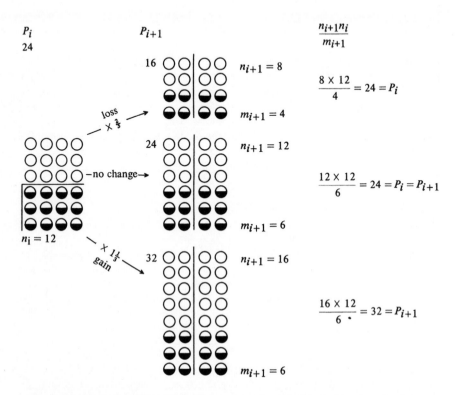

P_i
24

P_{i+1}

$$\frac{n_{i+1}n_i}{m_{i+1}}$$

16      $n_{i+1} = 8$

$$\frac{8 \times 12}{4} = 24 = P_i$$

loss $\times \frac{2}{3}$ $m_{i+1} = 4$

24      $n_{i+1} = 12$

-no change→

$$\frac{12 \times 12}{6} = 24 = P_i = P_{i+1}$$

$m_{i+1} = 6$

$n_i = 12$

$\times 1\frac{1}{3}$ gain   32 $n_{i+1} = 16$

$$\frac{16 \times 12}{6} = 32 = P_{i+1}$$

$m_{i+1} = 6$

Animals from the population on the left are captured, marked and released on day i. A second sample is taken on day $i + 1$ and a certain number of marked animals are recaptured. The right hand sets represent the population after loss, no change and gain respectively. Animals to the right of the vertical lines are included in the sample.

P_i, P_{i+1} population size on days i and $i + 1$

n_i size of sample taken day i, i.e. number of animals captured, marked and released day i

n_{i+1} size of second sample taken day $i + 1$

m_{i+1} number of recaptured (marked) animals in second sample

In the declining population the first day's sampling fraction is equal to the fraction of marks in the second sample:

$$\frac{n_i}{P_i} = \frac{m_{i+1}}{n_{i+1}} \quad \text{or} \quad P_i = \frac{n_{i+1}n_i}{m_{i+1}}$$

Figure 3.1 continued

In the population gaining individuals, the second day's sampling fraction is equal to the fraction marks recaptured/marks released:

$$\frac{n_{i+1}}{P_{i+1}} = \frac{m_{i+1}}{n_i} \quad \text{or} \quad P_{i+1} = \frac{n_{i+1}n_i}{m_{i+1}}$$

Both identities are true for the static population.

Note that the sampling fractions in all the above samples are the same (0·5) and therefore sampling effort on day $i + 1$ has increased from the declining to the gaining population. The method is still applicable when the sampling fraction does not remain constant.

---

in the second sample are marked and recaptured in a third sample, together with marks applied on the first occasion. Let us mark the animals in the first sample *red* and the animals in the second sample *blue*. Since we have a second sample containing both unmarked animals and animals with a red mark, and since we apply the blue mark of day $i + 1$ regardless of whether the animal has been previously marked, in the third sample we shall recapture animals with red marks, with blue marks and with both. For simplicity, exactly the same numbers are assumed to be captured on each day so that n_i, n_{i+1} and n_{i+2} are all equal. Let us further assume that all animals captured are successfully marked and released, so that numbers captured equal numbers released. Figure 3.2, on front and back cover, represents two examples of this triple catch.

On the back cover the set of three columns represents a population on three successive days in which there is gain. The set on the front cover represents a population losing individuals. From these populations a sample is taken on each of the three days; the first column of the trio represents the population just after the first sample has been released, the second and third depict the populations at the moments of sampling. The animals sampled are on the right. The sample is the same size on each of the three days ($n_i = n_{i+1} = n_{i+2} = 54$ back cover, 72 front cover). Animals in the samples are marked red on day i and blue on day $i + 1$; they are arranged in rows to ensure a fair proportion of marks in the sample.

A cursory inspection of the increasing population shows that the absolute number of red marks remains unaltered, but the ratio of reds to the total number of individuals in the sample changes. By contrast, in the declining population the absolute number of reds changes, but the ratio of reds to the total individuals in the sample remains unaltered. Note that only the individuals on the right of the vertical lines are known to the investigator.

(b) Analysis using number of marks

(i) Handling the data: trellis type 1

We can summarise the data from the triple catch illustrated in Figure 3.2 by means of the trellis diagram shown in Table 3.1. The daily catches are symbolised by the letters A, B and C representing n_i, n_{i+1} and n_{i+2}. The recaptures are symbolised by the appropriate small letter with a subscript indicating when they were recaptured. Thus a_1 and a_2 represent recaptures of A (animals marked red on day i) on days $i + 1$ and $i + 2$ and b_1 represents recaptures of B (animals marked blue on day $i + 1$) on day $i + 2$.

In each cell of the trellis a number of recaptured marks is entered. At the head of the column in which the cell is included we have the date the marks were applied; on the right of the row on which the cell lies we have the date of recapture. All marks are counted; remembering that some animals have two marks it is clear that there will be a larger number of marks entered than the number of marked animals. The trellis can be expanded for any number of catches.

(ii) Indicators of rates and signs of population change

Figure 3.2 and Table 3.1 show:
- *In the population gaining individuals:* The number of red marks remains the same but the proportion of reds in the population declines and this decline is revealed in the sample where $a_2 < a_1$. The blues and reds in sample $i + 2$ are equal in number, $b_1 = a_2$.

- *In the population losing individuals:* The number of red marks declines but the proportion of reds in the population remains the same, $a_2 = a_1$, there are more blues than reds in the sample $i + 2$, $b_1 > a_2$.
  A difference in the number of reds recaptured on days $i + 1$ and $i + 2$ indicates gain. A difference in the number of blue and red marks recaptured on day $i + 2$ indicates loss.

(iii) Rate of gain: relation between two successive recaptures of first day's mark (reds recaptured on days i + 1 *and* i + 2)

In the population losing individuals the ratio of red marks sampled at $i + 1$ and $i + 2$ must remain equal and since in the example $A = B = C$, the actual numbers of red marks in the two samples should be the same ($a_1 = a_2$). In the population gaining individuals, the fraction of red marks in sample $i + 2$ must be less than in sample $i + 1$ ($a_2 < a_1$) since the same number of a more dilute mixture is sampled at $i + 2$. The ratio of a_1 to a_2 gives the finite rate of growth of the population:

Table 3.1

Trellis type 1 listing marks.

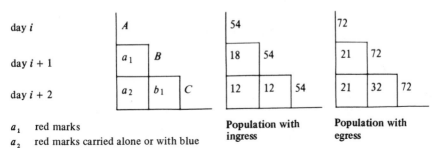

a_1	red marks		**Population with**	**Population with**
a_2	red marks carried alone or with blue		**ingress**	**egress**
b_1	blue marks carried alone or with red			

Each cell has co-ordinates of date of release (above) and date of capture (on right). The capital letters symbolise both numbers captured and released and the date-specific mark applied.

$$\lambda^+ = \frac{a_1}{a_2} \qquad (3.6)$$

The reciprocal of the finite rate is referred to as the positive survival rate:

$$s^+ = \frac{a_2}{a_1} \qquad (3.7)$$

and the rate of addition of animals to the population is $\lambda^+ - 1$.

(iv) Rate of loss: relation between recaptures on third day of two successive days markings (reds and blues applied days i and i + 1 and recaptured day i + 2)

In the population gaining individuals the numbers of red marks and of blue marks in the third sample (a_2 and b_1) will be equal since the population clearly cannot gain marked individuals. In the population losing individuals, there will be fewer red marks than blue in the third sample ($a_2 < b_1$) since the red marks will have been leaving over a period of two days, whereas the blue marks will have suffered only one day's loss. The ratio of red to blue marks recaptured on day $i + 2$ gives the finite rate of decline of the population, which is equivalent to the negative survival rate:

$$s^- = \frac{a_2}{b_1} \qquad (3.8)$$

The rate of loss of animals from the population is the complement of the negative survival rate, i.e. $1 - s^-$.

(v) Modification of the Lincoln Index to take account of loss and gain

Having recognised the existence of loss when $b_1 > a_2$ and gain when $a_1 > a_2$ we now know which of our two basic equations, (3.2) and (3.5), will give us the correct population estimate. We can summarise the appropriate population estimates as in Table 3.2. Notice that the left-hand expressions have the same form as Equation 3.2 and the right-hand expressions have the same form as Equation 3.5.

Table 3.2
Population estimation when there is either loss or gain but not both.

$\hat{P}$	Losing but not gaining	Gaining but not losing	
$P_A =$	$\dfrac{AB}{a_1}$		(3.9)
$P_B =$	$\dfrac{BC}{b_1}$	$\dfrac{AB}{a_1}$	(3.10)
$P_C =$		$\dfrac{BC}{b_1}$	(3.11)

Now let us consider a population suffering both loss and gain during each interval between samples. The simple Equations 3.2 and 3.5 will not be valid; gain will increase the value of B/a_1 in (3.9) and C/b_1 in (3.10); loss will increase the value of A/a_1 in (3.10) and B/b_1 in (3.11) so that all four estimates will be too high. However, the rates of change computed from the triple catch are still valid even for a population gaining and losing at the same time, and we can therefore correct the population estimates by multiplying them by the appropriate survival rates as in Table 3.3. Strictly speaking, s^+ and s^- estimate survival from

Table 3.3
Population estimations when there is both loss and gain.

$\hat{P}$	Correction for gain $\times s^+$		Correction for loss $\times s^-$	
$P_A =$	$\dfrac{AB}{a_1} \times \dfrac{a_2}{a_1}$			(3.12)
$P_B =$	$\dfrac{BC}{b_1} \times \dfrac{a_2}{a_1}$	$=$	$\dfrac{AB}{a_1} \times \dfrac{a_2}{b_1}$	(3.13)
$P_C =$			$\dfrac{BC}{b_1} \times \dfrac{a_2}{b_1}$	(3.14)

day $i + 1$ to day $i + 2$ and therefore the estimate for the population size on the first day may not be valid. The most interesting feature of our set of corrected estimates is the existence of two estimates for P_B; these two estimates have different logical derivations but are arithmetically identical (since $A = B = C$, but see Note 4.4(a)). The logic by which these estimates are derived is essentially that of Jackson (1939).

(c) Analyses using numbers of marked individuals

So far, only marks have been noticed, irrespective of the animals carrying them. In the third sample from the population with loss illustrated in Figure 3.2 there are 12 animals with red marks alone, 23 with blue marks alone and 9 with both red and blue marks together. In the trellis for this sample we entered 32 blue marks and 21 red marks. The fact that 9 animals had both marks has not been considered in our argument. Is this useful information which we have ignored?

(i) Handling the data: listing each individual according to the mark or combination of marks it carries

The three categories of animals in the third sample (3) have different histories which we can tabulate as in Table 3.4.

Table 3.4
Histories of marked individuals.

Animals in sample 3		Numbers of each category in sample 3 to nearest whole number	Presence or absence in sample		
			1	2	3
X	blue	23	—	+	+
Y	red and blue	9	+	+	+
Z	red	12	+	—	+

While our categories are first distinguished by the mark or marks they carry, we are really interested in what the marks tell us about the status of the animals that carry them; in particular, about their status in the second sample (2). Viewing each category as documented above, we may observe that in sample 2:

category *X* is present, has no known past, but a known future
category *Y* is present, has a known past and a known future
category *Z* is absent, but has a known past and a known future

Note that category Z, although absent from sample 2, must have been present in the population at the time since it is present in samples 1 and 3. Categories Y and Z both include animals with red marks and their status in past, present and future is known, whereas category X includes animals with blue marks alone and their status is known only in the present and future.

(ii) Trellis type 2

In the type 1 trellis we entered red marks as a_1 and a_2 and blue marks as b_1. In the type 2 trellis we enter *marked individuals*; those with red marks as a_1 and a_2, with blue marks as b_1 and with both red and blue marks as **ab**. Note that a_1 is the same in both trellises. For the remainder of Chapter 3, note that a distinction is made between **bold** and *italic* letters.

(iii) Analysis using categories Y and Z

Amongst the sample of animals marked red on day i and recaptured on day $i + 2$ we know that category Z (a_2) were available in the population on day $i + 1$ but were not captured; category Y animals (ab) were also available for recapture since they were, in fact, captured. We can therefore estimate the sample fraction on day $i + 1$ from these red-marked animals recaptured on day $i + 2$ as follows:

$$SF_B = \frac{\text{red-marked animals recaptured on}}{\text{red-marked animals known to be available}} = \frac{ab}{ab + a_2} \qquad (3.15)$$
$$\text{days } i + 1 \text{ and } i + 2$$
$$\text{for recapture on day } i + 1$$

We have chosen to ignore animals with blue marks and those animals which had red marks but did not appear in sample 3. The identity in Equation 3.4 depends on n_i, the number of red marks applied, suffering no loss between i and $i + 1$ and therefore the right-hand term cannot be used without the application of a correction factor for loss. Now Equation 3.15 gives an estimate of SF which is not affected by loss. There may be loss, but the categories **ab** and a_2 are clearly not lost since they reappear in sample 3. Therefore we can write:

$$SF_B = \frac{ab}{ab + a_2} \qquad (3.16)$$

$$P_B = \frac{B(ab + a_2)}{ab} \qquad (3.17)$$

This is the method of Manly and Parr (1968, see Ch. 4).

Jackson (1948) saw the possibilities of this re-recapture : recapture ratio, but the full realisation of its potential is due to Jolly (1965), Seber (1965) and Manly and Parr (1968).

(iv) Analysis using category Z

Recall the Equation 3.5 on page 30 for the population on the second day of a double catch:

$$P_{i+1} = \frac{n_i n_{i+1}}{m_{i+1}}$$

This identity depends on all marks on day i surviving to day $i+1$ and being available for recapture. For a true estimate, n_i needs correcting by multiplying it by the survival rate, s^-. The product $n_i s^-$ gives an estimate of the marks available for recapture on the next occasion; we call this $\hat{M}_{i+1}$. We can replace the n_i in the equation above by this corrected estimate, bringing all the subscripts the same:

$$\hat{P}_{i+1} = \frac{n_{i+1} \hat{M}_{i+1}}{m_{i+1}} \tag{3.18}$$

$\hat{M}_{i+1}$, or specifically for our triple catch, $\hat{M}_2$, is the number of red-marked animals available for recapture on day $i+1$; it is initially an unknown quantity. However, we do know that a_1 of the $\hat{M}_2$ red-marked animals were recaptured on day $i+1$ and therefore $\hat{M}_2 - a_1$ were available but not caught. Of these $\hat{M}_2 - a_1$ animals eluding capture on day $i+1$, a_2 (the Z category) were recaptured on day $i+2$. We also know that, of the blue-marked animals, B, b_1 ($= ab + b_1$) were recaptured on day $i+2$. The two statements above allow us to attach probabilities to the risk of capture immediately after the marked animals from sample 2 had been released:

Probability of red-marked animals absent from sample 2 being captured immediately afterwards	$\dfrac{a_2}{\hat{M}_2 - a_1}$
Probability of blue-marked animals (with or without red) being captured immediately afterwards	$\dfrac{b_1 + ab}{B}$

These two probabilities should be equal; the risk of capture should be the same for all marks and, indeed, all unmarked animals. We can therefore equate the two and solve for $\hat{M}_2$:

$$\frac{a_2}{\hat{M}_2 - a_1} = \frac{b_1 + ab}{B} \tag{3.19}$$

therefore

$$\hat{M}_2 = \frac{Ba_2}{b_1 + ab} + a_1 \tag{3.20}$$

We can now substitute this value for $\hat{M}_2$ in Equation 3.18 to give the population estimate:

$$\hat{P}_2 = \frac{B}{a_1} \left(\frac{Ba_2}{b_1 + ab} + a_1 \right) \tag{3.21}$$

This is the method of Jolly (1965) and Seber (1965). The right hand term of (3.19) may be rewritten as m_{i+1}/n_i which is the estimate of SF_{i+1} on the right of Equation 3.4 which is only valid when there is no loss. In (3.19) it is the risk of capture immediately after sample 2 was taken and released; that is to say, it is the risk before loss from the population has had time to operate.

(v) Analysis using the numbers of animals carrying only the mark of the second day

Using a similar argument to that used by Jolly we can derive the probabilities of recapture of two different categories of marked animals and equate these, to estimate $\hat{M}_2$. Consider the newly captured animals on day $i + 1$ to which a blue mark is applied $(B - a_1)$; of these, b_1 are recaptured on day $i + 2$. Consider secondly the red *marks* applied on day i; of these, an unknown number $\hat{M}_2$ will survive to be available for recapture on day $i + 1$ and a_2 will be recaptured on day $i + 2$. The fractions $b_1/B - a_1$ and $a_2/\hat{M}_2$ are probabilities of recapture and should both be equal to the sampling fraction and therefore to each other:

$$SF_2 = \frac{b_1}{B - a_1} = \frac{a_2}{\hat{M}_2}$$

therefore

$$\hat{M}_2 = \frac{a_2(B - a_1)}{b_1} \tag{3.22}$$

substituting this value for $\hat{M}_2$ in the simple Lincoln Index:

$$\hat{P}_2 = \frac{\hat{M}_2 B}{a_1}$$

we get:

$$\hat{P}_2 = \frac{a_2(B - a_1)B}{a_1 b_1} \tag{3.23}$$

This is Bailey's triple catch method (Bailey 1951).

Table 3.5
Trellis type 2 listing marked animals.

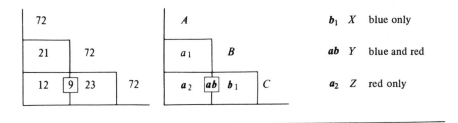

72				*A*		b_1 *X*	blue only
21	72			a_1	*B*	*ab* *Y*	blue and red
12	9	23	72	a_2 *ab* b_1	*C*	a_2 *Z*	red only

(d) The triple catch: a synthesis and conclusions

The formulae for the methods of analysis of data from a triple catch are listed in Table 3.6. The symbols used in the second and third columns are those explained in Tables 3.1 and 3.5; and it is important to discriminate between the bold italicised letters representing marked animals and the italicised letters representing total marks.

In column 3 the first two formulae have larger numerators and denominators than the second two and we can therefore expect the latter pair to be more susceptible to sampling error. Notice that the denominators of the first two formulae representing the methods of Jackson, Fisher & Ford and Jolly are the same, that of Bailey is slightly different involving the term *b*, (the animals with blue marks alone) whereas the denominator of Manly and Parr is the single value *ab*, the double-marked animals which will be the smallest category of recaptures; this must mean that the method of Manly and Parr will be the most susceptible to sample size.

We can learn most by observing the similarities and differences between the formulae (1) and (2) and between (3) and (4). Formulae (1) and (2) have $B^2 a_2/a_1 b_1$ in common whilst *ab* in (1) is replaced by $a_1 b_1/B$ in (2): (3) and (4) have Ba_2 in common, *ab* in (3) being replaced by $a_1 b_1/B - a_1$ in (4). The terms not held in common are the observed and expected number of the double-marked animals, *ab*. The expected value of *ab* in Jolly's formula can be explained in terms of simple probability as follows:

probability of red recaptures on day B = a_1/B
probability of blue recaptures on day B = b_1/C
(i.e. immediately after release)

therefore probability of red and blue =

$$\frac{a_1 b_1}{BC}$$

Therefore expected number of animals with red and blue marks on day C =

$$\frac{a_1 b_1}{BC} C = \frac{a_1 b_1}{B}$$

The expected value of **ab** in Bailey's formula can be derived as follows. Immediately after release on day $i + 1$ there are a_1 animals marked both red and

---

Table 3.6

Formulae for estimating the size of a population from a triple catch.

General formulae (used in Ch.4)	Formulae for triple catch (used in this chapter)	Formulae modified for comparison	
Jackson, Fisher & Ford			
$\hat{P}_i = \dfrac{n_i \hat{M}_i}{m_i}$	$\hat{P}_B = \dfrac{B^2 a_2}{a_1 b_1}$	$\dfrac{B^2(a_2 + ab)}{a_1 b_1}$	(1)
$\hat{M}_i = n_{i-1}s$			
Jolly			
$\hat{P}_i = \dfrac{n_i \hat{M}_i}{m_i}$	$\hat{P}_B = \dfrac{B}{a_1}\left[\dfrac{B a_2}{b_1 + ab} + a_1\right]$	$\dfrac{B^2\left(a_2 + \dfrac{a_1 b_1}{B}\right)}{a_1 b_1}$	(2)
$\hat{M}_i = \dfrac{n_i Z_i}{R_i} + m_i$			
Manly and Parr			
$\hat{P}_i = \dfrac{n_i(Y_i + Z_i)}{Y_i}$	$\hat{P}_B = \dfrac{B(ab + a_2)}{ab}$	$\dfrac{B a_2}{ab}$	(3)
Bailey			
	$\hat{P}_B = \dfrac{B(B - a_1)a_2}{a_1 b_1}$	$\dfrac{B(B - a_1)a_2}{a_1 b_1}$	(4)

See Note 4.4 for proof that correction for differing sample sizes is not necessary for Jackson, and that Jackson and Fisher and Ford are identical when there is a triple catch.

blue of which **ab** are recaptured on day $i + 2$, and there are $B - a_1$ animals which have only a blue mark of which b_1 are recaptured on day $i + 2$. Assuming the probability of an animal being caught on day $i + 2$ is the same, whichever mark or combinations of marks it carries, we can therefore write:

$$\frac{ab}{a_1} = \frac{b_1}{B - a_1}$$

and:

$$ab = \frac{a_1 b_1}{B - a_1}$$

In conclusion, we must stress that only in the triple catch situation do Jackson's positive and negative estimates for day $i + 1$ give an identical estimate and only here is the Fisher and Ford estimate identical to that of Jackson.

None of the methods of analysis, except that of Bailey, were designed for use with only three sampling occasions. In an extended MRR programme the rarer categories Y and Z tend to grow at the expense of the X category and all the advantages of larger values in the sample fractions accrue. The principles derived from examination of the model with reference to a triple catch apply equally to any number of sampling occasions; the additional detail necessary to process data from an extended MRR sequence is given in the next chapter.

4 MRR methods in detail

Introduction

We have shown that the principal methods of analysis of MRR data fall into two categories: those which use the number of marks regardless of whether they occur singly or in multiples, and those which use the number of marked individuals and utilise the information from multiple recaptures. Accordingly, we begin by explaining the handling of MRR data from an actual class exercise. There follows a section on each of the methods, illustrated by this worked example.

Handling the data

(a) The field book table

This is shown on the left in Figure 4.1. Each recaptured individual is represented by a row; these cross as many columns as sampling occasions. Data for the trellis type 1 are read off directly from the sub-totals.

Manly and Parr (1968) devised a modification of this field book table in order to gather together in convenient form information on the different combinations present in each row.

(b) The Manly and Parr table

Entries in this table are similar to those in the field book table, but now a stroke appears for each animal each time it is captured; an extra column of strokes for the last day is added (see Figure 4.1).

Some animals appear on more than one day of the field book table; in the Manly and Parr table no animal appears twice; construction of the table therefore involves finding those animals common to more than one day and grouping all such entries into a single row of strokes. This grouping can be done progressively from day to day or left until the end of the programme of sampling and done retrospectively. The former method is suited to studies with a high proportion of recaptures; the latter is useful when few recaptures are obtained.

In addition to assembling animals into categories X, Y and Z the Manly and Parr table serves two other functions:

(i) It assembles date-specific marks into individual specific combinations of marks. It is possible to apply individually specific marks in the first place but this can be time consuming and technically difficult. A Manly and Parr table gives us the same amount of information about the frequency of animals with each kind of history using single date-specific marks.

(ii) It exposes discrepancies of diagnosis and marking. Suppose we found two individuals on day F carrying marks B, C and D. We know that only one animal with such a mark combination exists in the field. It would follow that something has gone wrong with the scoring system, a mark had been applied to one category (species, stage or sex) and recorded in the field book table under another category. The Manly and Parr table, when constructed progressively, keeps a running check on errors of this kind and for this reason alone it is preferable to compile the table from day to day by the progressive method.

Figure 4.1
Field book table and Manly and Parr table.

The left side shows the day-to-day entry of information as it is collected, the right side is the Manly and Parr table constructed as follows. The first day's captures are entered as strokes in column A. Recaptures on day B are added against five of the first day individuals, the 28 new captures being recorded in a separate column. On day C, five animals with marks of day A are entered against the next five first day captures, 12 animals with the mark of day B against the first 12 day B individuals and the remaining $26 - 5 - 12 = 9$ are entered as new animals beneath the previous entries. As we proceed, the space accorded to newly captured animals is likely to decline as the detail for each of the other animals grows. Finally, we alter all strokes preceded and succeeded by strokes to Y's, any spaces left between strokes become Z's, and the unmodified strokes represent the X category. A similar method may be used to construct a table by the retrospective method.

The total number of animals in each of categories Y and Z are determined for each day from all columns except the first and last. Animals in these categories for day i are known to be present on previous and subsequent occasions. Animals in category Z are known to be available for recapture on day i but do not, in fact, appear in the sample for that day.

This example represents real data obtained by sampling on consecutive days a population of fourth instar nymphs of the grasshopper *Myrmeleotettix maculatus*, on short grassland in southern England. A square of ground was searched, of 60 m edge, divided into four equally sized small squares. The insects were caught by hand and marked with spots of cellulose paint on the pronotum. The data have been analysed by all the methods discussed, and the results are given in Notes 4.1 to 4.6.

Figure 4.1 continued

Day	Captures		Recaptures						Manly and Parr table					
	Total	New	A	B	C	D	E	F	A	B	C	D	E	F
A	20	卌 卌 卌 卌							/	y	z	/		
B	33	卌	/						/	/				
		卌	/						/	/				
		卌	/						/	/				
		卌	/						/	/				
		卌	/						/	z	y	/		
		/// (28)	5						/	z	y	/		
C	26	卌	/						/	z	y	z	z	/
		//// (9)	/						/	z	/			
			/						/	z	y	z	/	
			/						/	z	z	y	/	
				/					/					
				/					/					
				/					/	× 7				
				/					/	y	y	/		
				/					/	y	y	z	/	
				/					/	y	/			
			+7						/	y	/			
			5 12						/	y	/			
D	28	卌	/						/	/				
		卌		/					/	/				
		// (12)		/					/	/				
				/					/	/				
				/					/	/				
					/				/	/				
					/				/	z	y	y	/	
			/	/	·				/	z	y	z	/	
			/	·	/				/	z	/	z		
			/	·	/				/	z	/			
			·	/	/				/	z	z	/		
			·	/	/				/	z	z	/		
			·	/	/				/	z	z	/		
			·	/	/				/	z	z	/		
			·	/	/				/	z	z	/		
			4 11 9						/	z	z	/		

Figure 4.1 continued

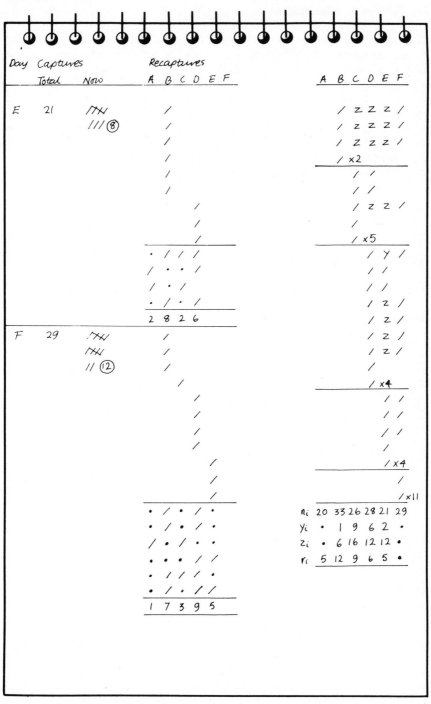

(c) The two types of trellis

We have already used two types of trellis (Tables 3.1 and 3.5) in which a_1, a_2 and b_1 are the number of marks recaptured, a_2 and b_1 are the number of individuals recaptured with either mark A or mark B, and ab is the number of animals re-captured with both marks A and B. Remember the capital letters indicate both date and number of captures and releases; each recapture figure has its date of release above and its date of capture to the right.

If we extend our sampling to a fourth occasion we find no difficulty in extending our type 1 trellis, but what about our type 2 trellis? The little box idea to include the multiple recaptures becomes cumbersome. The numbers of combinations of four marks on the fifth day is 11, i.e. a total of 15 categories to be housed on the row marked E. For n marks we have $2^n - 1$ marks and combinations of marks; for example, for eight sampling occasions we have 7 marks to

---

Table 4.1

Type 1 and type 2 trellises extended to four days.

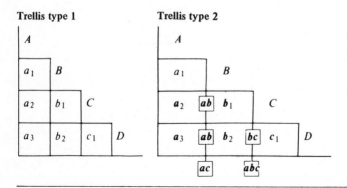

Table 4.2

Jolly's modification of the type 2 trellis.

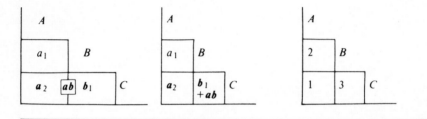

---

apply giving 127 marks and combinations on the eighth occasion. Jolly's modification of trellis type 2 (Table 4.2, cf. Table 3.5) is to eliminate the box giving separate mention of the individuals marked **ab** and to combine them with those marked **b**. In the Jolly trellis, each individual is entered in the column of its most recent mark. This means that once a mark has appeared in the type 2 trellis, it does not appear again in that column, whereas in the type 1 trellis, all marks are entered every time they appear in combination. From this point we shall not use bold letters for the entries in the trellis type 2.

In an extended Jolly trellis we can distinguish the entries in the right-hand cell of each row (the cells containing entries with the subscript $_1$) as the *hypotenuse cells* from those to the left, to which we can refer as the *internal cells* (containing entries with subscripts $_2$ or higher). Entries in the hypotenuse cells represent animals present both on the date of recapture and on the previous day (these are the r_i animals of Manly and Parr, Note 4.6). It follows that these

---

Table 4.3

Two categories of cells in the Jolly trellis.

Hypotenuse cells open
Internal cells cross hatched

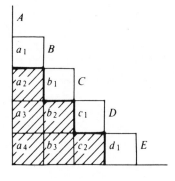

---

animals cannot be regarded as Z's on either their release or recapture date. Now the entries occupying the internal cells were *not* captured on the occasion immediately following their release; they were absent on the occasion before their recapture and are, therefore, the Z animals for each of the days between their release and recapture. The geometry of the situation in relation to the Manly and Parr table is presented in Table 4.4 which summarises everything we have said about the Manly and Parr table and the modified type 2 trellis of Jolly.

In Note 4.1 we show how all essential raw data may be incorporated into trellis form as a useful check on the arithmetic operations.

Table 4.4
Relation between trellis type 2 of Jolly and the Manly and Parr type entries.

Day recaptured	Internal cell	Possible mark combinations	M & P notation	Hypotenuse cell	Possible mark combinations	M & P notation
C	a_2	A	$X\ Z\ X$	b_1	B	$X\ X$
					AB	$X\ Y\ X$
D	a_3	A	$X\ Z\ Z\ X$	c_1	C	$X\ X$
	b_2	AB	$X\ Y\ Z\ X$		AC	$X\ Z\ Y\ X$
		B	$X\ Z\ X$		BC	$X\ Y\ X$
					ABC	$X\ Y\ Y\ X$
E	a_4	A	$X\ Z\ Z\ Z\ X$	d_1	D	$X\ X$
	b_3	B	$X\ Z\ Z\ X$		AD	$X\ Z\ Z\ Y\ X$
		AB	$X\ Y\ Z\ Z\ X$		BD	$X\ Z\ Y\ X$
	c_2	C	$X\ Z\ X$		CD	$X\ Y\ X$
		AC	$X\ Z\ Y\ Z\ X$		ABD	$X\ Y\ Z\ Y\ X$
		BC	$X\ Y\ Z\ X$		ACD	$X\ Z\ Y\ Y\ X$
		ABC	$X\ Y\ Y\ Z\ X$		BCD	$X\ Y\ Y\ X$
					$ABCD$	$X\ Y\ Y\ Y\ X$

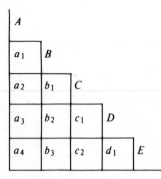

Note 4.1

Worked example. All data incorporated in trellis form.

Trellis types 1 and 2(a) are the two which we have just described (chapter 4(c)). A further two types 2(b) and 2(c) enable us to store all the necessary information contained in the Manly and Parr table; this is a useful device for cross-checking the arithmetic.

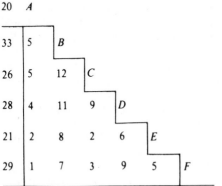

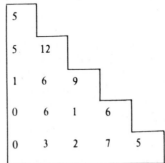

Trellis type 1

Enter all marks, whether held singly or in combinations for Jackson and Fisher and Ford.

Trellis type 2(a)

Enter marked animals, according to the date of the most recent mark for Jolly and Manly and Parr. Internal cells are Z animals, hypotenuse cells are r_i animals for M & P.

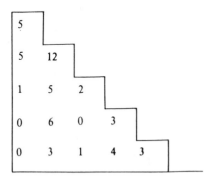

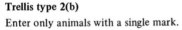

Trellis type 2(b)

Enter only animals with a single mark.

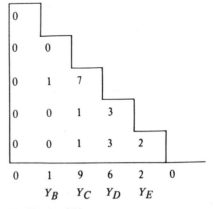

Trellis type 2(c)

Subtract entries in 2(b) from 2(a) columns summed give Y_i for M & P.

Jackson's method

In a type 1 trellis we have columns of successive recaptures (small letters) of a mark applied which is indicated at the top of the column. The capital letter at the top of the column indicates both the date of release and the type of mark applied. Horizontally, reading from right to left, we have rows of recaptures of successive days markings, captured on the date indicated by the capital letter on the right. In the second numerical example (Fig. 3.2) we had the same number of captures on each day; we also assumed that all these captured animals were released. We shall continue to assume the latter for ease of description and notation, but not the former; it is not always possible in practice to capture the same number on each occasion. For the purpose of Jackson's analysis we correct the recapture figures in the body of the trellis to match an agreed standard number, X, captured and released. Correction is a matter of simple proportion; the method will be evident from Table 4.5.

We have established, in the triple catch, that $a_2 \leqslant a_1$ and that a difference between the two must be due to gain; similarly, we showed that $a_2 \leqslant b_1$, any difference being due to loss from the population (see pp. 34, 35). By the same arguments any recapture figure in the body of the trellis must be less than, or equal to the figure above and to the right. In a population subject to change, figures in columns will decline towards the base; figures in rows will decline to the left (Table 4.5). Any figure divided by the one above gives the survival rate of marks subject to dilution because of gain to the population (s^+). Any figure divided by the one to the right gives the survival rate of marks subject to loss from the population between successive markings (s^-). In Chapter 3 we used these survival rates to correct the population estimate by the simple Lincoln Index (Table 3.3). This is the essence of Jackson's method.

(a) Smoothing for sampling error

(i) Positive series: columns

Consider the left-hand column of the trellis where we have a history of marks applied the first day. Remember that all recapture figures in the trellis now represent the fraction of marks in the population.

$$\text{Survival of marks } A, \text{ from } B \text{ to } C \ = \ \frac{a_2}{a_1}$$

$$\text{from } C \text{ to } D \ = \ \frac{a_3}{a_2}$$

$$\text{from } D \text{ to } E \ = \ \frac{a_4}{a_3}$$

Table 4.5

Jackson's method. Trellis type 1 correction of recaptures and indications of gain and loss.

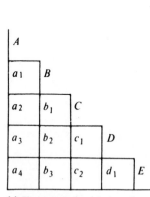

(a) **Uncorrected capture and recapture figures**

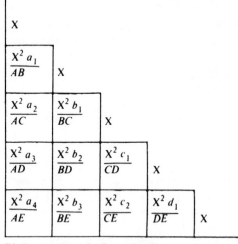

(b) **Corrected entries for each cell**

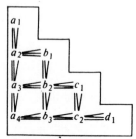

(c) **Relation between corrected recapture figures when there is loss and gain**

← s^+ decline due to gain

↓ s^- decline due to loss

If the population changes are smooth and the sampling perfect we can write:

$$s^+ = \frac{a_2}{a_1} = \frac{a_3}{a_2} = \frac{a_4}{a_3} \tag{4.1}$$

but the sampling will not be regular and some kind of averaging is necessary. We could take the arithmetic mean of the three fractions, although Jackson suggests:

$$s^+ = \frac{a_2 + a_3 + a_4}{a_1 + a_2 + a_3} \tag{4.2}$$

Another possibility would be the geometric mean.

(ii) Negative series: rows

Consider the lowest row; here we have the figures of successive markings re-captured on the last day. Ideally we can write:

$$s^- = \frac{a_4}{b_3} = \frac{b_3}{c_2} = \frac{c_2}{d_1} \tag{4.3}$$

and the equivalent average is:

$$s^- = \frac{a_4 + b_3 + c_2}{b_3 + c_2 + d_1} \tag{4.4}$$

(b) Population estimates: first method

Jackson used the positive survival rate from the first column to extrapolate back to a hypothetical figure a_0 representing the recaptures which would have been obtained if the second sample had been taken immediately following the release of the first. This is theoretically acceptable but practically impossible since no time would be allowed for the marked animals to intersperse with the unmarked; however, it is a useful way of looking at the matter.

As we outlined in Chapter 3, we need only correct the simple Lincoln Index by multiplying by the appropriate survival rate:

$$P_A = \frac{X^2 s^+}{a_1} \tag{4.5}$$

$$P_E = \frac{X^2 s^-}{d_1} \tag{4.6}$$

(c) Second method

If we accept the geometrical relation between the recapture figures in the trellis, remembering that any pair of figures in a column will give a figure for s^+ and any pair in a row will give a figure for s^-, we can smooth between the first two columns as follows:

$$s^-_{A \to B} = \frac{a_2 + a_3 + a_4}{b_1 + b_2 + b_3} \qquad (4.7)$$

We can also smooth between the last two rows:

$$s^+_{D \to E} = \frac{c_2 + b_3 + a_4}{c_1 + b_2 + a_3} \qquad (4.8)$$

Note that all the entries in the top row of (4.7) and (4.8) refer to a single day and those in the bottom row refer to the previous day, so that the survival rate measures the change between the two days. The population estimates are then:

$$P_B = \frac{X^2 s^-_{A \to B}}{a_1} \qquad (4.9)$$

$$P_D = \frac{X^2 s^+_{D \to E}}{d_1} \qquad (4.10)$$

The first method can be applied to any column or row, or combination of columns or rows, the second to any adjacent pair of columns or rows. The amount of information in the columns and rows declines to the right and towards the upper part of the trellis, but with more sampling occasions (a larger trellis), population estimates and survival rates can be derived from more columns and rows. The first method is appropriate to one marking occasion followed by successive sampling without further marking. Survival rates from the second method can be used to test the desirability of using a constant smoothed rate for the first method, or for Fisher and Ford's method. An example based on real data from a grasshopper population is given in Note 4.2.

We have mentioned previously (Table 3.6) that the method of Jackson and the following method of Fisher and Ford become identical when simplified to a triple catch situation. An explanation of the simplification and proof of the identity are given in Note 4.4.

Note 4.2

(a) Worked example: Jackson's method (see Table 4.5).

Entries in the trellis are adjusted to a standard value by the method described in Table 4.5.

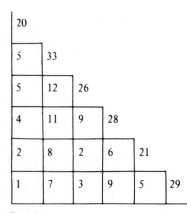

Raw data:

20					
5	33				
5	12	26			
4	11	9	28		
2	8	2	6	21	
1	7	3	9	5	29

Adjusted data:

20					
3·03	20				
3·85	5·59	20			
2·86	4·76	4·95	20		
1·90	4·62	1·47	4·08	20	
0·69	2·93	1·59	4·43	3·28	20

Survival rates

First Method

Positive series, all data:

$$s^{+} = \frac{0{\cdot}69 + 1{\cdot}90 + 2{\cdot}86 + 3{\cdot}85 + 2{\cdot}93 + 4{\cdot}62 + 4{\cdot}76 + 1{\cdot}59 + 1{\cdot}47 + 4{\cdot}43}{1{\cdot}90 + 2{\cdot}86 + 3{\cdot}85 + 3{\cdot}03 + 4{\cdot}62 + 4{\cdot}76 + 5{\cdot}59 + 1{\cdot}47 + 4{\cdot}95 + 4{\cdot}08}$$

$$= 0{\cdot}784$$

Negative series, all data:

$$s^{-} = \frac{0{\cdot}69 + 2{\cdot}93 + 1{\cdot}59 + 4{\cdot}43 + 1{\cdot}90 + 4{\cdot}62 + 1{\cdot}47 + 2{\cdot}86 + 4{\cdot}76 + 3{\cdot}85}{2{\cdot}93 + 1{\cdot}59 + 4{\cdot}43 + 3{\cdot}28 + 4{\cdot}62 + 1{\cdot}47 + 4{\cdot}08 + 4{\cdot}76 + 4{\cdot}95 + 5{\cdot}59}$$

$$= 0{\cdot}772$$

Second Method

Positive series, between days:

$$s^{+}_{C \to D} = \frac{2{\cdot}86 + 4{\cdot}76}{3{\cdot}85 + 5{\cdot}59} = 0{\cdot}807$$

$$s^{+}_{D \to E} = \frac{1{\cdot}90 + 4{\cdot}62 + 1{\cdot}47}{2{\cdot}86 + 4{\cdot}76 + 4{\cdot}95} = 0{\cdot}636$$

$$s^{+}_{E \to F} = \frac{0{\cdot}69 + 2{\cdot}93 + 1{\cdot}59 + 4{\cdot}43}{1{\cdot}90 + 4{\cdot}62 + 1{\cdot}47 + 4{\cdot}08} = 0{\cdot}799$$

Negative series, between days:

$$s^{-}_{A \to B} = \frac{0{\cdot}69 + 1{\cdot}90 + 2{\cdot}86 + 3{\cdot}85}{2{\cdot}93 + 4{\cdot}62 + 4{\cdot}76 + 5{\cdot}59} = 0{\cdot}520$$

$$s^{-}_{B \to C} = \frac{2{\cdot}93 + 4{\cdot}62 + 4{\cdot}76}{1{\cdot}59 + 1{\cdot}47 + 4{\cdot}95} = 1{\cdot}537$$

$$s^{-}_{C \to D} = \frac{1{\cdot}59 + 1{\cdot}47}{4{\cdot}43 + 4{\cdot}08} = 0{\cdot}360$$

Note 4.2 continued

Population estimates

Day	Survival over all days s^-	Survival between days	
	$P_{i+1} = \dfrac{X^2 s^-}{n_i}$	$P_i = \dfrac{X^2 s^+_{i \to i+1}}{n_i}$	$P_{i+1} = \dfrac{X^2 s^-_{i \to i+1}}{n_i}$
B	$\dfrac{400 \times 0{\cdot}772}{3{\cdot}03} = 102$		$\dfrac{400 \times 0{\cdot}520}{3{\cdot}03} = 69$
C	$\dfrac{400 \times 0{\cdot}772}{5{\cdot}59} = 55$	$\dfrac{400 \times 0{\cdot}807}{4{\cdot}95} = 65$	$\dfrac{400 \times 1{\cdot}537}{5{\cdot}59} = 110$
D	$\dfrac{400 \times 0{\cdot}772}{4{\cdot}95} = 62$	$\dfrac{400 \times 0{\cdot}636}{4{\cdot}08} = 62$	$\dfrac{400 \times 0{\cdot}360}{4{\cdot}95} = 29$
E	$\dfrac{400 \times 0{\cdot}772}{4{\cdot}08} = 76$	$\dfrac{400 \times 0{\cdot}799}{3{\cdot}28} = 97$	
F	$\dfrac{400 \times 0{\cdot}772}{3{\cdot}28} = 94$		

A very similar sequence of population estimates is obtained using s^+ over all days, but these should be credited to one day earlier. Values of s^+ or s^- greater than unity are theoretically impossible and arise from sampling error.

If s^- can be taken to be constant then the number entering the population between i and $i+1$ is $g_{i \to i+1} = P_{i+1} - P_i s^-$.

(b) Worked Example: Simple Lincoln Index and Jackson's triple catch.

The Lincoln Index gives an estimate of P_i when there is loss but no gain, and P_{i+1} when there is gain but no loss. The triple catch method cannot provide estimates for the first or last days.

Day	Lincoln Index	Triple catch
A		
	$20 \times 33/5 = 132$	
B		$33^2 \times 5/(5 \times 12) = 91$
	$33 \times 26/12 = 72$	
C		$26^2 \times 11/(12 \times 9) = 69$
	$26 \times 28/9 = 81$	
D		$28^2 \times 2/(9 \times 6) = 29$
	$28 \times 21/6 = 98$	
E		$21^2 \times 9/6 \times 5) = 132$
	$21 \times 29/5 = 122$	
F		

Fisher and Ford's method

Instead of deriving a survival rate directly from the type 1 trellis as in Jackson's method, Fisher and Ford use an arbitrarily chosen rate from which they calculate the number of marks expected to survive. The rate is validated by comparison with the observed number of days survived by recaptured marks, or rejected and a new value tried.

(a) Validating the survival rate

The first step is to sum the entries in the rows of the trellis to give the total number of marks recaptured each day (Table 4.6, column 1). In the next column (column 2) each mark is counted for each day it has survived to give the total marks x days.

Table 4.7 shows a similar trellis but this time including all the marks expected to survive in the population on a given day, and available for recapture. These entries are derived as the products of the numbers of animals released, A, B, C, etc; and successive powers of the survival rate. The rate is s^-.

In column 4 we have 'marks times days'; if we divide this by column 3, the result is 'days survived per mark' (column 5). The product of 'days survived per mark' and actual recaptures (column 1) is the expected number of mark/days survived. This value in column 6 is then compared with that in column 2. The value given to s^- is then changed and the process repeated until the difference between observed and expected is reduced to an acceptably low level. A practical method for carrying out this operation is given in Chapter 6 (p. 99).

(b) The population estimate

Having established s, the value in column 3 will now be a good estimate of the marks available for recapture on the given day, $\hat{M}_i$. The complementary value in column 1 gives us the marks recaptured on the given day, m_i. $\hat{M}_i$ and m_i are now substituted in the equation:

$$\hat{P}_i = \frac{n_i \hat{M}_i}{m_i} \qquad (4.11)$$

where $\hat{M}_i$ = entry in column 3 for day i

$\quad m_i$ = entry in column 1 for day i and

$\quad n_i$ = the number of releases on day $i, A, B, C, \ldots$ etc: (cf. Table 3.6)

The estimates for the grasshopper population are given in Note 4.3.

Table 4.6

First step in Fisher and Ford's method.

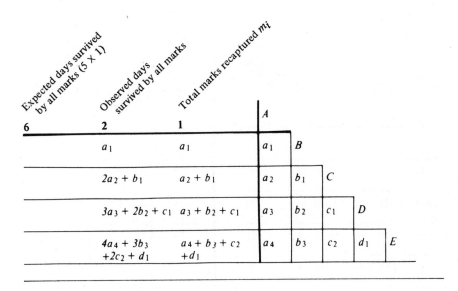

Expected days survived by all marks (5×1)	Observed days survived by all marks	Total marks recaptured m_i	A	B	C	D	E
6	2	1	A				
	a_1	a_1	a_1	B			
	$2a_2 + b_1$	$a_2 + b_1$	a_2	b_1	C		
	$3a_3 + 2b_2 + c_1$	$a_3 + b_2 + c_1$	a_3	b_2	c_1	D	
	$4a_4 + 3b_3 + 2c_2 + d_1$	$a_4 + b_3 + c_2 + d_1$	a_4	b_3	c_2	d_1	E

Table 4.7

Second step in Fisher and Ford's method.

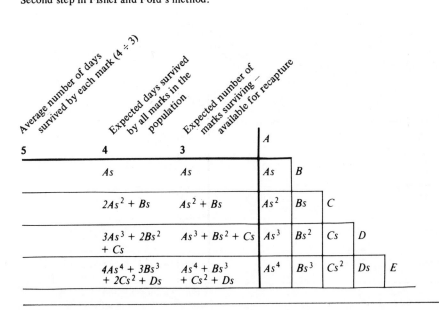

Average number of days survived by each mark ($4 \div 3$)	Expected days survived by all marks in the population	Expected number of marks surviving – available for recapture	A	B	C	D	E
5	4	3	A				
	As	As	As	B			
	$2As^2 + Bs$	$As^2 + Bs$	As^2	Bs	C		
	$3As^3 + 2Bs^2 + Cs$	$As^3 + Bs^2 + Cs$	As^3	Bs^2	Cs	D	
	$4As^4 + 3Bs^3 + 2Cs^2 + Ds$	$As^4 + Bs^3 + Cs^2 + Ds$	As^4	Bs^3	Cs^2	Ds	E

Note 4.3

Worked Example: Fisher and Ford's Method (cf. Tables 4.6 and 4.7).

Trellis type 1 recaptured marks

$(2-6)$	Expected days survived by all marks (5×1)	Observed days survived by all marks	Total marks recaptured m_i	A	B	C	D	E	F
7	6	2	1						
0	$1 \cdot 0 \times 5 = 5$	5	5	5					
$-0 \cdot 61$	$1 \cdot 33 \times 17 = 22 \cdot 61$	22	17	5	12				
$-0 \cdot 44$	$1 \cdot 81 \times 24 = 43 \cdot 44$	43	24	4	11	9			
$2 \cdot 58$	$2 \cdot 19 \times 18 = 39 \cdot 42$	42	18	2	8	2	6		
$-1 \cdot 50$	$2 \cdot 66 \times 25 = 66 \cdot 50$	65	25	1	7	3	9	5	

Average days survived by each mark $(4 \div 3)$	Total days survived by all marks T_i	Total marks surviving available for recapture $\dot{M}_i$	20					
5	4	3						
1	$16 \cdot 26$	$16 \cdot 26$	$16 \cdot 26$	33				
$1 \cdot 33$	$53 \cdot 27$	$40 \cdot 05$	$13 \cdot 22$	$26 \cdot 83$	26			
$1 \cdot 81$	$97 \cdot 00$	$53 \cdot 70$	$10 \cdot 75$	$21 \cdot 81$	$21 \cdot 14$	28		
$2 \cdot 19$	$145 \cdot 28$	$66 \cdot 42$	$8 \cdot 74$	$17 \cdot 73$	$17 \cdot 19$	$22 \cdot 76$	21	
$2 \cdot 66$	$189 \cdot 18$	$71 \cdot 07$	$7 \cdot 10$	$14 \cdot 42$	$13 \cdot 97$	$18 \cdot 51$	$17 \cdot 07$	29

Value of s chosen $= 0 \cdot 813$, giving a difference of $0 \cdot 03$ in column 7.

Note 4.3 continued

Calculation of entries in columns 3 and 4 may be performed in a sequential manner using the equations:

$$M_i = s(M_{i-1} + n_{i-1}) \qquad M_0 = 0$$

$$T_i = sT_{i-1} + M_i$$

Population estimates

Day	$P_i = \dfrac{n_i \hat{M}_i}{m_i}$
B	$\dfrac{33 \times 16\cdot26}{5} = 107$
C	$\dfrac{26 \times 40\cdot05}{17} = 61$
D	$\dfrac{28 \times 53\cdot70}{24} = 63$
E	$\dfrac{21 \times 66\cdot42}{18} = 77$
F	$\dfrac{29 \times 71\cdot07}{25} = 82$

Note 4.4

(a) The special case of Jackson's triple catch.

In our presentation of Jackson's method it has been assumed that A, B and C are identical or that a_1, a_2 and b_1 have been corrected for equal values of A, B and C. For the analysis of a triple catch by Jackson's method we do not need to apply a correction to the recapture figures when A, B and C are not equal. Consider the correction factors which would be applied:

$$
\begin{array}{|c|c|c|}
\hline
X & & \\
\hline
\dfrac{X^2 a_1}{AB} & X & \\
\hline
\dfrac{X^2 a_2}{AC} & \dfrac{X^2 b_1}{BC} & X \\
\hline
\end{array}
$$

and substitute these in Equation 3.13:

$$P_B = X^2 \frac{X^2 a_2}{AC} \cdot \frac{AB}{X^2 a_1} \cdot \frac{BC}{X^2 b_1}$$

the two X^2's top and bottom cancel, as also the A's and C's leaving:

$$P_B = \frac{B^2 a_2}{a_1 b_1}$$

Estimates for a series of days are shown in Note 4.2(b) together with the Lincoln Index estimates. These may be compared with the results in Note 4.2(a).

(b) The identity of Jackson and Fisher & Ford in a triple catch.

Using the symbols of Fisher and Ford's calculation for the triple catch calculation (see Tables 4.6 and 4.7), we have:

Expected (5×1)	Observed			
6	2	1	A	
a_1	a_1	a_1	a_1	B
$\dfrac{2As^2 + Bs}{As^2 + Bs}(a_2 + b_1)$	$2a_2 + b_1$	$a_2 + b_1$	a_2	b_1	C

First step

Note 4.4 continued

$5(4 \div 3)$	4	3	A		
1	As	As	As	B	
$\dfrac{2As^2 + Bs}{As^2 + Bs}$	$2As^2 + Bs$	$As^2 + Bs$	As^2	Bs	C

Second step

The survival rate s is calculated by equating the observed and expected days survived by all marks (*column 6* cf. *column 2*). For more than three days this is done by minimising the difference. When there are three days:

$$\frac{2As^2 + Bs}{As^2 + Bs}(a_2 \div b_1) = 2a_2 + b_1$$

therefore

$$(2As^2 + Bs)(a_2 + b_1) = (As^2 + Bs)(2a_2 + b_1)$$

from which it can be shown that:

$$s = \frac{Ba_2}{Ab_1}$$

Substitute this value for s in the general equation for P:

$$
\begin{aligned}
P_B &= \frac{AB}{a_1}s \\
&= \frac{AB}{a_1}\frac{Ba_2}{Ab_1} \\
&= \frac{B^2a_2}{a_1b_1}
\end{aligned}
$$

which is identical to Equation 3.13, Table 3.3.

Jolly's method

This method, developed independently by Jolly (1965) and Seber (1965) uses the information available in statistically the most efficient way. In principle, it is therefore the best method to use, but later we discuss the merits of some of the other methods in particular conditions.

(a) Principle

The basic equation used is the simple Lincoln Index of the second form (Equation 3.5) which is not affected by gain but must be corrected for loss. Like Fisher and Ford, the correction for loss is incorporated into a term $\hat{M}_i$, but this term is not the number of marks but the number of *marked animals* surviving from all previous releases. The method uses the type 2 trellis (see Table 4.2) which lists recaptures of animals according to their most recent mark. The estimate of the number of marked animals available for recapture, $\hat{M}_i$, is obtained in a different way. The value is incorporated as an unknown into an equation which is then solved for this value. The equation uses two new terms R_i and Z_i derived from the modified type 2 trellis shown in Table 4.8.

   R_i is the total of recaptured animals occupying one column of the trellis; i.e. animals marked on day i and recaptured subsequently.
   Z_i is the total number of animals marked on a day previous to i, not caught on day i, but caught subsequently.

By summing the rows of the type 2 trellis from the left (right-hand trellis in Table 4.8) the Z_i animals are accumulated in the columns and the m_i animals are summed in the hypotenuse cells.

(b) Explanation: population estimate

Consider the marked animals in the population just after the sample of day i. First there are the n_i animals which have just been released. Second there are the older marked animals which were not captured on day i. This second category is the unknown number of animals, $\hat{M}_i$, available for recapture on day i minus those which were recaptured on day i, m_i, which have now joined the first category, n_i.
   The most recently marked animals, n_i, and the animals carrying older marks, $\hat{M}_i - m_i$, are at risk of recapture immediately after the sample of day i. Of the newly marked n_i animals, we know that R_i will eventually be recaptured; of the older $\hat{M}_i - m_i$ animals, we know that Z_i were subsequently recaptured.
   It is a basic assumption of the Lincoln Index that marked animals behave in

Table 4.8

The Jolly trellis and variables derived from it.

hypotenuse cells
internal cells

Trellis type 2(a) (Jolly)   **Trellis for Z_i**

sum rows left to right ⟶ hypotenuse cells now contain total recaptures

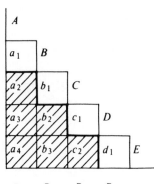   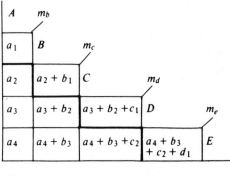

R_A R_B R_C R_D Z_B Z_C Z_D

sum all entries in columns   sum columns of internal cells only
to give sums of successive
recaptures of animals released
on a given day

every respect like the unmarked. It follows that any category of marked animals should behave like any other category, in particular, the n_i and the $\hat{M}_i - m_i$ animals should be equally at risk any time after the sample of day i. We can therefore equate the fractions of the n_i and $\hat{M}_i - m_i$ animals which were re-captured:

$$\frac{R_i}{n_i} = \frac{Z_i}{\hat{M}_i - m_i} \tag{4.12}$$

and solve for $\hat{M}_i$

$$\hat{M}_i = \frac{n_i Z_i}{R_i} + m_i \tag{4.13}$$

and finally substitute this value in the basic equation:

$$\hat{P}_i = \frac{n_i \hat{M}_i}{m_i} \tag{4.14}$$

Equation 4.12 will still be true whatever the variations in sampling intensity from day $i + 1$ to the end of the programme.

(c) Survival rate

Equation 4.12 is however, dependent on the newly marked category n_i surviving at the same rate as the older category, $\hat{M}_i - m_i$. Making the assumption of an age-independent survival, Jolly estimates the rate as follows:

Number of marked animals in the population immediately
after sample i $= \hat{M}_i - m_i + n_i$
Number of these which survive to day $i + 1$ $= \hat{M}_{i+1}$

therefore

$$\hat{M}_{i+1} = s_{i \to i+1} (\hat{M}_i - m_i + n_i)$$

and

$$s_{i \to i+1} = \frac{\hat{M}_{i+1}}{\hat{M}_i - m_i + n_i} \tag{4.15}$$

The numerical example is given in Note 4.5 where the method of estimating gain is also evident.

Note 4.5
Worked Example: Jolly's method (cf. Table 4.8).

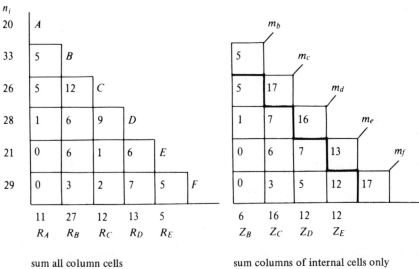

Left trellis:

n_i						
20	A					
33	5	B				
26	5	12	C			
28	1	6	9	D		
21	0	6	1	6	E	
29	0	3	2	7	5	F
	11	27	12	13	5	
	R_A	R_B	R_C	R_D	R_E	

sum all column cells
Trellis type 2(a)

Right trellis (Trellis for Z_i):

	m_b	m_c	m_d	m_e	m_f
5					
5	17				
1	7	16			
0	6	7	13		
0	3	5	12	17	
6	16	12	12		
Z_B	Z_C	Z_D	Z_E		

sum columns of internal cells only
Trellis for Z_i

Population estimates

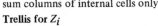

Day	$\hat{M}_i = \dfrac{Z_i n_i}{R_i} + m_i$	$P_i = \dfrac{n_i \hat{M}_i}{m_i}$
A	$= 0$	
B	$\dfrac{6 \times 33}{27} + 5 = 12\cdot33$	$\dfrac{33 \times 12\cdot33}{5} = 81$
C	$\dfrac{16 \times 26}{12} + 17 = 51\cdot66$	$\dfrac{26 \times 51\cdot66}{17} = 79$
D	$\dfrac{12 \times 28}{13} + 16 = 41\cdot85$	$\dfrac{28 \times 41\cdot85}{16} = 73$
E	$\dfrac{12 \times 21}{5} + 13 = 63\cdot40$	$\dfrac{21 \times 63\cdot40}{13} = 102$

Note 4.5 continued

Survival and gain

Interval	Survival rate	Number gained
	$$s_{i \to i+1} = \dfrac{\hat{M}_{i+1}}{\hat{M}_i - m_1 + n_i}$$	$$g_{i \to i+1} = \hat{P}_{i+1} - \hat{P}_i s_{i \to i+1}$$
$A-B$	$\dfrac{12 \cdot 33}{0 - 0 + 20} = 0 \cdot 617$	
$B-C$	$\dfrac{51 \cdot 66}{12 \cdot 33 - 5 + 33} = 1 \cdot 281$	$79 \cdot 0 - (81 \cdot 4 \times 1 \cdot 281) = -25 \cdot 3$
$C-D$	$\dfrac{41 \cdot 85}{51 \cdot 66 - 17 + 26} = 0 \cdot 690$	$73 \cdot 2 - (79 \cdot 0 \times 0 \cdot 690) = 18 \cdot 7$
$D-E$	$\dfrac{63 \cdot 40}{41 \cdot 85 - 16 + 28} = 1 \cdot 177$	$102 \cdot 4 - (73 \cdot 2 \times 1 \cdot 177) = 16 \cdot 2$

Survival rates greater than unity are theoretically impossible and gain should not be estimated from such rates. The calculation is set out to show the method. The negative gain results from an aberrant survival rate and does not indicate loss. These remarks apply equally in Note 4.6.

Manly and Parr's method

Like Jolly, Manly and Parr make use of the Z category. In addition, they use a new category, Y.

Z_i animals are the marked individuals known to be present on day i, but not recaptured.
Y_i animals are those recaptured on day i which were also captured both before and after day i.

Values for Z_i and Y_i are available directly from the Manly and Parr table (see Fig. 4.1). Alternatively (or in addition, for cross checking) Z_i may be taken from trellis type 2(a) as extended for Jolly's method (Note 4.5) and Y_i is available in trellis type 2(c) (see Note 4.1).

(a) Explanation: population estimate

In addition to the categories Z_i and Y_i, the only other category of marked animal in the population is the X category represented by unmodified strokes in the Manly and Parr table (see Fig. 4.1), either on day i or on a previous day. The probability of any category of marked animal being recaptured on day i is the same and is equal to the sample fraction, SF_i. Manly and Parr derive an estimate of this using only the Y and Z categories and ignoring the X category:

$$\hat{SF_i} = \frac{\text{Marked animals with past and future known, recaptured on day } i}{\text{Marked animals with past and future known, available for recapture on day } i} = \frac{Y_i}{Y_i + Z_i}$$

(4.16)

Manly and Parr use their estimate of SF_i directly to derive their estimate of population size as follows:

$$\hat{P_i} = \frac{n_i}{\hat{SF_i}}$$

$$= \frac{n_i(Y_i + Z_i)}{Y_i}$$

(4.17)

(b) Survival rate

Manly and Parr's estimate of $\hat{P_i}$ does not assume that any marked animal, whenever the mark was applied, has an equal probability of surviving to be recaptured. Running down a Manly and Parr table it will be evident that there are roughly equal numbers of y and z entries before and after day i. The Manly and Parr estimate is therefore freer from the assumption of an age-independent survival rate. This freedom is gained by using a smaller proportion of the marks available in the population and the method is, therefore, more dependent on a high sampling intensity than Jolly's estimate.

The survival rate is estimated using the value r_i which is derived from the Manly and Parr table or from the hypotenuse cells of the trellis type 2(a) (see Note 4.1); r_i is the number of animals present in the sample on day i and also present on day $i + 1$.

$s_{i \rightarrow i+1} n_i$ animals should survive to day $i + 1$
$s_{i \rightarrow i+1} n_i SF_{i+1}$ animals should be recaptured on day $i + 1$

In fact, r_i animals *are* recaptured on day $i + 1$, therefore:

$$r_i = s_{i \rightarrow i+1} n_i SF_{i+1}$$

therefore

$$s_{i \rightarrow i+1} = \frac{r_i}{n_i SF_{i+1}}$$

(4.18)

The numerical example is in Note 4.6.

Note 4.6

Worked example: Manly and Parr's method.

Summations from Manly and Parr table (Figure 4.1).

	A	B	C	D	E	F
n_i	20	33	26	28	21	29
Y_i		1	9	6	2	
Z_i		6	16	12	12	
r_i	5	12	9	6	5	

Population estimates

Day	$SF_i = \dfrac{Y_i}{Y_i + Z_i}$	$\hat{P}_i = \dfrac{n_i}{SF_i}$
B	$\dfrac{1}{7}$	$\dfrac{33 \times 7}{1} = 231$
C	$\dfrac{9}{25}$	$\dfrac{26 \times 25}{9} = 72$
D	$\dfrac{6}{18}$	$\dfrac{28 \times 18}{6} = 84$
E	$\dfrac{2}{14}$	$\dfrac{21 \times 14}{2} = 147$

Survival and gain

Interval	Survival rate $s_{i \to i+1} = \dfrac{r_i}{n_i \, SF_{i+1}}$	number gained $g_{i \to i+1} = \hat{P}_{i+1} - \hat{P}_i s_{i \to i+1}$
A–B	$\dfrac{5}{20} \times \dfrac{7}{1} = 1{\cdot}750$	
B–C	$\dfrac{12}{33} \times \dfrac{25}{9} = 1{\cdot}010$	$72 - (231 \times 1{\cdot}010) = -161$
C–D	$\dfrac{9}{26} \times \dfrac{18}{6} = 1{\cdot}038$	$84 - (72 \times 1{\cdot}038) = 9$
D–E	$\dfrac{6}{28} \times \dfrac{14}{2} = 1{\cdot}500$	$147 - (84 \times 1{\cdot}500) = 21$

(c) Temporary emigration

The methods of Jolly and Manly and Parr rely on an estimate of the numbers of individuals available in the population but uncaught, Z_i. Such animals are assumed to be present because they were caught both before and after day i. In reality, animals may leave the area, to return later, so that uncaught animals are not necessarily available for capture. This temporary migration may lead to inaccurate estimates; the area to which the population is ascribed may need redefining. All the MRR methods may be affected by this phenomenon of temporary absence but the danger happens to be highlighted by the two methods which use the Z category in making population estimates.

Separating death from emigration

In the grasshopper example (see Fig. 4.1) individuals may be lost by death, ecdysis into the next instar, or emigration. The first two processes cannot be separated, but a method exists for separating loss *in situ* from emigration. The total sample square of 60×60 m was divided up into four squares each measuring 30×30 m and, by applying the mark to different parts of the pronotum, the insects were given marks specific to the square in which they were caught. In the calculations a recapture is scored as such regardless of whether or not it was found in the square where it was first marked. However, it would be possible to regard a recapture taken outside the square in which it was marked as an emigrant and ignore it, retaining records only of those individuals trapped and retrapped in the same small square. Data for the total area compiled in this way could then be used to estimate the survival rate.

Now the ratio of edge to area for the small squares is twice that for the large square. We should therefore expect a lower estimate of survival from the small-square data, due to the greater emigration recognised when the vagrant recaptures are ignored. If d represents the loss rate *in situ* and e represents the emigration rate from the large square, then:

$$1 - s_{\text{large square}} = d + e$$

and:

$$1 - s_{\text{small square}} = d + 2e$$

These equations may be solved to give estimates of d and e. The survival rate for the large square by the Fisher and Ford method was 0.813 (Note 4.3); for the small squares it was 0.780. Using these values we find d to be 0.154 and e to be 0.033. In practice, large samples are required, so that accurate estimates of s may be made, to ensure reasonable reliability. The method was devised by Jackson (1939) who also used small and large square estimates of his s^+ to separate birth from immigration.

Minimum number of animals known to be alive

Some studies of populations of small rodents have used a technique for estimating the minimum number of animals alive, sometimes called the direct census. The total is obtained by adding the number sampled n_i and the number known to be present but not captured Z_i. The result cannot include an estimate of the animals that are never caught, as do methods derived from the Lincoln Index, and hence will underestimate P to an unknown and varying degree. Furthermore, no allowance is made for the effect of loss. The example provides values of 39, 42, 40 and 33 for days B to E. These may be compared with the estimates in Table 7.2. This and other inadequacies of the method are revealed by the computer simulations of Hilborn *et al.* (1976). Many of the problems of estimating the size of rodent populations arise from the method by which samples are obtained; a population is trapped repeatedly during a restricted interval of time. Under these conditions trap avoidance and trap addiction may both be exhibited in the population (Bishop and Hartley 1976). These problems can be overcome at least partially by sampling less frequently (Ch. 7, p. 109) or by using a frequency of capture method (Ch. 5). They are not surmounted by direct census methods.

Expectation of life

In all the methods except that of Manly and Parr survival is assumed to be constant throughout the life of an individual, and the survival rate s is estimated per unit time (per day). If this assumption holds then the mean expectation of life of an individual is $1/(1 - s)$, when the survival rate is high. If s is estimated as 0.9 the average length of life is 10 days. As s declines towards zero so this value, which is based on a discontinuous drop in numbers, declines to one. A correction for continuity may therefore be made by reducing the expectation of life by a half. More accurately, on the assumption of exponential decline in numbers, the life expectancy may be calculated as:

$$e_0 = \frac{-1}{\ln s} \tag{4.19}$$

Population geneticists are frequently interested in survival rates as measures of fitness. If the age-specific fecundity of two forms in a population is constant and the same, but they survive at different rates, then the relative fitness of the forms is the ratio of their expectations of life. Bishop, Cook and Muggleton (1978) obtained estimates of the daily survival rate in a rural area of a melanic form and the non-melanic typical form of the moth *Biston betularia.* Using the Fisher and Ford method they were 0.28 and 0.49 respectively. From Equation 4.19 the expectations of life are 0.79 and 1.41 days, so that the fitness of the typical relative to the melanic is 1.41/0.79 or 1.79, an advantage of 79%. When the Jolly or Manly and Parr methods are used the geometric mean may be calculated from a series of individual estimates.

Many insects have annual life cycles, so that the total population of adults in a given generation may be investigated. Fisher and Ford (1947) required this estimate when studying genetic change from one generation to the next in a population of tiger moths. If a series of k estimates of the daily population size, P_i, are obtained throughout the season, then the total population emerging is:

$$\frac{1}{e_0} \sum_{i=1}^{i=k-1} P_i + P_k.$$

5 Time samples with constant sampling effort

Introduction

Several methods have been developed to estimate P from numbers or ratios in successive samples. They are usually applicable when both loss from and gain to the population are negligible during the sampling but the animals are mobile. They thus have a characteristic of area samples — the closed nature of the population — and of MRR — the mobility of the individuals. Some of them make use of data where the animals are caught and removed.

As an example of one type of data which may be obtained, consider the following results for four successive sampling sessions of the Small Heath butterfly, *Coenonympha pamphilus*, made during a student exercise in southern England. They were actually intended to be the first sample of a mark, release and recapture census, but heavy rain the following day led to a very small second sample. Nevertheless, some information can be obtained about the population by considering the results as a trapping problem. Two students searched one of four adjacent squares for half an hour, catching all the butterflies they could see, recording their number and putting them into a large keep net to be released after sampling was complete. After half an hour each pair moved round to one of the other squares and continued for a further half hour. In this way the total area was searched for a total of two hours, a constant effort being put into each successive half hour. The population was unlikely to change during this period as a result of gain or loss. The resulting totals were 167 for the first period, 90 for the second, 76 for the third and 65 for the fourth.

Constant catching effort is the first requirement of the methods to be described. We will assume that each individual has the same probability of capture at one time, and a constant probability of capture from period to period.

Trapping and removal

(a) Two sample method

The numbers in the four successive samples discussed in the previous section drop progressively. If the fall is due to the removal of individuals from the population, then P may be estimated. Call the samples n_1, n_2, n_3 and n_4. Then the sampling intensity, or probability of capture, may be estimated as $p = n_1/P$. If this probability remains constant then another estimate of it is $n_2/(P - n_1)$. This leads to the equation:

$$\frac{n_1}{P} = \frac{n_2}{P - n_1}$$

so that:

$$\hat{P} = \frac{n_1^2}{n_1 - n_2} \qquad (5.1)$$

Only two samples are therefore needed to estimate P. The three values provided by the successive pairs n_1 and n_2, n_2 and n_3, and n_3 and n_4 are 362, 579 and 525.

(b) Several successive samples, maximum likelihood method

Since the value P being estimated is the same for each pair a method is needed to combine all the data available to produce a single estimate of the population size. Two approaches to this problem will be considered. The first, called the method of maximum likelihood, is derived from the probabilities of drawing s successive samples of size $n_1, n_2, n_3, \ldots n_s$ from a population of size P. If we consider one sample, the probability of getting n_1 individuals is $C\,p^{n_1} q^{P-n_1}$, where p is the probability of catching an animal, q is the probability of it not being caught ($= 1 - p$) and C is an integer representing the number of ways in which n_1 individuals may be picked from the population (see Ch. 6, p. 92). After the first sample has been taken the population consists of $P - n_1$ individuals, so the probability of getting n_2 in the second sample is $C'\,p^{n_2} q^{P-n_1-n_2}$. The probability of getting both n_1 and n_2 in successive samplings is the product of these two expressions, and for s samples the probability is a compound of s such terms. The problem is to find the values of p and q which give the greatest likelihood of getting the observed samples $n_1, n_2, \ldots n_s$. This is done by ignoring the C terms, differentiating the logarithm of the rest with respect to q and equating the resulting expression to zero in order to maximise q.

This approach was used by Moran (1951) and developed by Zippin (1956). Two equations are needed, namely:

$$\frac{sq^s}{1 - q^s} - \frac{q}{1 - q} + \frac{\Sigma(i - 1)n_i}{\Sigma n_i} = 0 \qquad (5.2)$$

from which q is estimated, and:

$$\hat{P} = \frac{\Sigma n_i}{1 - q^s} \qquad (5.3)$$

In words, $\hat{P}$ is the total number caught divided by the probability of being captured on one or more of the s occasions. Equation 5.2 may be solved by trial and error, which presents no difficulty on a calculator, although the process is speeded up on a programmable machine. The calculations are shown in Note 5.1.

Equations 5.2 and 5.3 reduce to Equation 5.1 for the two-sample case. They are an efficient means of finding the sampling intensity, and hence the population size, for a sequence of any number of samples.

Note 5.1

Trapping and removal: steps in the calculation of the maximum likelihood method.

Four successive samples were taken of the butterfly *Coenonympha pamphilus*. The effort put into catching each sample was the same and the numbers caught declined progressively. The data were:

$n_1 = 167 \qquad s = 4$
$n_2 = 90$
$n_3 = 76$
$n_4 = 65$

From these figures we calculate:

$$\frac{\Sigma(i-1)n_i}{\Sigma n_i} = \frac{437}{398} = 1\cdot098$$

Let $\qquad y = \dfrac{sq^s}{1-q^s} - \dfrac{q}{1-q} + \dfrac{\Sigma(i-1)n_i}{\Sigma n_i}$

Then by repeated trial and error:

q	y
0·70	0·029
0·71	0·013
0·72	−0·003

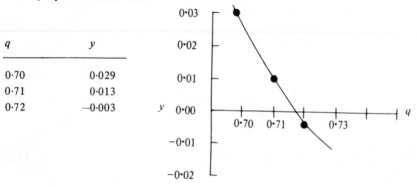

Note 5.1 continued

therefore,

$q = 0.72$ to an acceptable degree of accuracy, and:

$$1 - q^s = 0.7313$$

$$\hat{P} = \frac{\Sigma n_i}{1 - q^s} = \frac{398}{0.7313} = 544$$

In this example q is quite large and the accuracy of the estimate correspondingly low; q should preferably be 0.5 or less.

Figure 5.1

Exponential decline in numbers with trapping occasion for a closed population when the probability of capture remains constant.

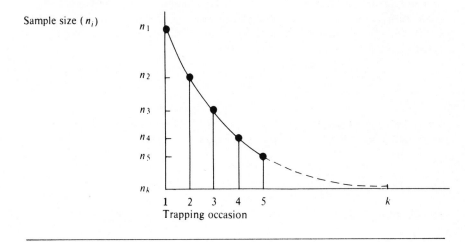

(c) Several samples, regression method

The basis of the argument outlined above is that the sampling fraction, or probability of capture, remains constant. If the probability of capture is p then $n_1 = Pp$. The remaining population is $P - Pp = P(1 - p)$, and $n_2 = Pp(1 - p)$. Let $q = (1 - p)$ as before. Consequently:

$$\frac{n_2}{n_1} = \frac{Pp(1 - p)}{Pp} = q \qquad (5.4)$$

Similarly, for any value i, n_{i+1}/n_i is an estimate of q. The curve of values of n_i against trapping occasion therefore declines by a constant fraction from step to step, and the value at which it has fallen to less than one is called n_k (Fig. 5.1). We only have the first few points on this curve. P may be estimated if the curve can be extrapolated to k so that we can sum all the values of n from 1 to k, since $P = \Sigma_{i=1}^{i=k} n_i$.

Such an extrapolation could be made in several ways. One method which follows from the assumption of constant sampling efficiency is to consider the relation of n_i to the sum of the values of n up to the one preceding the i'th capture (Fig. 5.2). Here we have n_1 plotted on zero, n_2 on n_1, n_3 on $n_1 + n_2$ etc. The decline in the ordinate from one step to the next is $n_j - n_{j+1}$, which from Equation 5.4 is equal to $n_j - qn_j$. The corresponding change in the abscissa is $\Sigma_{i=1}^{i=j-1} n_i - \Sigma_{i=1}^{i=j} n_i = -n_j$. Consequently, the slope of the line from one step to the next is always:

$$\frac{n_j(1-q)}{-n_j} = -p \qquad (5.5)$$

The line fitting the points is a straight one with gradient $-p$. If it is extrapolated to reach a value $n_k = 0$ the point at which it cuts the abscissa is an estimate of P.

The equation of this line is $n_i = c - px_i$, where x_i represents the successive cumulative values on the abscissa. The intercept c is the value of n for which $x = 0$, and the line passes through the mean point $\bar{x}, \bar{y}$. The slope may be estimated as the least squares regression coefficient:

$$-p = \frac{\Sigma(n_i - \bar{n})(x_i - \bar{x})}{\Sigma(x_i - \bar{x})^2}$$

where the summation is carried out over the s observations in the sample. Algebraically, this may be shown to be the same as:

$$-p = \frac{\Sigma n_i(x_i - \bar{x})}{\Sigma(x_i - \bar{x})^2} \qquad (5.6)$$

By rearranging the equation it is seen that the value of x for which $n = 0$ is:

$$\hat{P} = \bar{x} + \frac{\bar{n}}{p} \qquad (5.7)$$

The method was used by Leslie and Davis (1939) when estimating the number of rats in houses in West Africa from trapping records. Using the data for the butterflies we obtain $p = 0.3105$ from Equation 5.6 and $\bar{n} = 99.5$, $\bar{x} = 189.25$. The population estimate from Equation 5.7 is therefore 510. The calculation is shown in Note 5.2.

Note 5.2
Trapping and removal: steps in the calculation of the regression method.

The data for the butterfly *Coenonympha pamphilus* considered in Note 5.1 provide:

n_i	x_i	$s = 4$	
167	0	$\Sigma x = 757$	$\Sigma n = 398$
90	167	$\bar{x} = 189 \cdot 25$	$\bar{n} = 99 \cdot 5$
76	257	$\Sigma x^2 = 204827$	$\Sigma n^2 = 45990$
65	333	$\Sigma nx = 56207$	

The regression coefficient of Equation 5.6 is most easily calculated as:

$$-p = \frac{\Sigma nx - \dfrac{\Sigma n \Sigma x}{s}}{\Sigma x^2 - \dfrac{(\Sigma x)^2}{s}}$$

$$= \frac{-19114 \cdot 5}{61564 \cdot 75} = -0 \cdot 3105$$

Then:

$$\hat{P} = \bar{x} + \frac{\bar{n}}{p}$$

$$= 189 \cdot 25 + \frac{99 \cdot 5}{0 \cdot 3105}$$

$$= 509 \cdot 7$$

The graph of the relation between n and cumulative sample size is shown below:

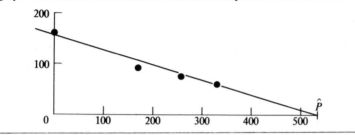

Increase in the fraction marked in a mark and release sequence

(a) Regression method

A similar method may be used when samples are marked and returned to the population. The requirements are not only as before, that the population should be closed with a constant probability of capture, but also that the capturing and marking process should leave an individual just as likely to be captured in the future as an unmarked one.

Suppose a sequence of samples is taken $- n_1, n_2, n_3, n_4$, etc. to n_s. All the individuals in each sample are marked and returned to the population. The whole of n_1 was unmarked on capture, but n_2 contains a number m_2 of individuals marked previously, n_3 contains m_3 marked individuals, and so on. As sampling continues, so the fraction m_i/n_i should increase as more and more of the population is marked. This fact may be made use of, in a method similar to the previous one, by examining the relation between the fraction marked in the i'th sample and the cumulative unmarked sample up to the $(i-1)$'th sample. Call the fraction marked y_i and the cumulative total unmarked x_i. Then $y_i = m_i/n_i$ and $x_i = \Sigma_{j=0}^{j=i-1}(n_j - m_j)$. As before, the relation of y_i to x_i is linear. The slope is positive and the line must pass through the origin $y_1 = x_1 = 0$ since no animals in a sample can bear a mark when none has previously been caught. We therefore have the equation:

$$y_i = bx_i \qquad (5.8)$$

where b is the slope. Note 5.3 and the argument in the previous section show that x_i is an estimate of P when $y_i = 1$. Consequently:

$$\hat{P} = \frac{1}{b} \qquad (5.9)$$

The estimate is therefore obtained if the slope of the line can be established. Setting $\bar{x} = 0$ and $\bar{y} = 0$ so that the line passes through the origin, the slope becomes:

$$b = \frac{\Sigma x_i y_i}{\Sigma x_i^2}$$

However, all the values of y_i are fractions of seemingly equal weight, whereas chance variation is likely to affect the sample sizes n_i on which they are based. The accuracy of y_i increases with increased sample size, and this may be allowed for by multiplying the squares and products by the sample size at time i. When this is done, and the regression equation inverted to provide the estimate of P, we get:

$$\hat{P} = \frac{\Sigma n_i x_i^2}{\Sigma n_i x_i y_i}$$

$$= \frac{\Sigma n_i x_i^2}{\Sigma m_i x_i} \tag{5.10}$$

Figure 5.2

Relation between sample size at the i'th capture and cumulative sample to the occasion preceding the i'th capture, for a closed population with constant probability of capture. $\hat{P}$ is the value x_k for which $n_k = 0$.

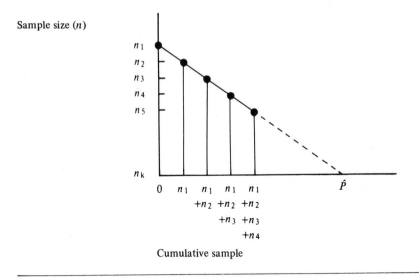

Sample size (n)

Cumulative sample

The method was devised by Schumacher and Eschmeyer (1943) to estimate fish populations in lakes, and was applied to small mammals by Hayne (1949, see also Caughley 1977a). In Note 5.3 an example is given using a population of the Bank vole *Clethrionomys glareolus*. The animals were caught in drop-fronted traps, marked so that each had a unique mark, and released. The uniqueness of the mark is not made use of in this estimation, for which it is simply necessary to know whether or not an individual has previously been caught. Over the period of the study the population was effectively a closed one, but of course the estimate will be affected if animals die or visitors arrive from outside the trapping area, or if capture rates of marked and unmarked animals differ.

(b) Maximum likelihood method

A maximum likelihood method is also available for data on frequency of recapture. When a probability function is set up in a manner analogous to that outlined on page 75, it may be shown (Seber 1973 p. 130) that:

$$1 - \frac{r}{P} - \prod_{i=1}^{i=s}\left(1 - \frac{n_i}{P}\right) = 0 \tag{5.11}$$

where s is the number of sampling occasions, n_i are the numbers in the successive samples and r is the total number of different animals caught. Π signifies the product of the s values within the parentheses. This equation may be solved by trial and error with successive values of P inserted. For the *Clethrionomys* data it gives $P = 61$.

When catching birds by netting, it may not be possible to take a sample, release it and then resample on a subsequent day. However, if the birds are caught, marked by banding and released immediately after capture, then information on population size is available from the frequency with which recaptures occur within the same trapping period.

The situation is an extreme version of the increase-in-fraction-marked method, in which the successive captures consist of one individual each. The total sample each time is either a marked recapture or an unmarked individual, so that the frequency of recapture events is considered, rather than the fraction of recaptures in each sample. The calculation of P may be carried out by the regression method, with $n_i = 1$ on all occasions and $m_i = 1$ or zero. Alternatively, with single captures, Equation 5.11 reduces to:

$$1 - \frac{r}{P} - \left(1 - \frac{1}{P}\right)^{s} = 0$$

This may be solved as before to give the population size. We are grateful to Dr M. V. Hounsome for drawing our attention to this application, which may be especially useful in bird-netting studies.

Note 5.3
Increase in fraction marked: regression method.

A population of the Bank vole *Clethrionomys glareolus* was sampled on six successive nights using a grid of drop fronted live-traps. The sample sizes (n_i) and numbers marked in each sample are shown in the second and third columns of the table below. The other columns give the figures required to estimate the population size and show the fraction marked on each occasion.

i	Sample size n_i	Number marked m_i	New $z_i = n_i - m_i$	Total new before i x_i	$n_i x_i^2$	$m_i x_i$	Fraction marked y_i
1	9	0	9	0	0	0	0·0
2	19	6	13	9	1539	54	0·32
3	18	10	8	22	8712	220	0·56
4	23	10	13	30	20700	300	0·43
5	34	26	8	43	62866	1118	0·76
6	8	6	2	51	20808	306	0·75
					114625	**1998**	

$$\hat{P} = \frac{\Sigma n_i x_i^2}{\Sigma m_i x_i} = \frac{114625}{1998} = 57\cdot4$$

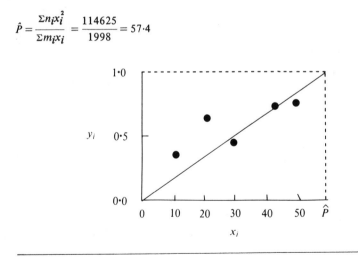

Estimation from frequency of recapture

When there are unique marks or different marks on each trapping occasion, so that it is known how many times each one has been recaptured, P can be estimated from the frequency of recapture of different individuals. If the chance of capture is random and small, then the probability of being captured once, twice, three times, etc. will follow a Poisson series, as shown in Chapter 2 (p. 22). Applying the Poisson distribution to the present situation we see that the first term is missing, since the number of animals not caught at all is unknown. The truncated series remaining may, therefore, be written:

number of times caught	1	2	3	... x
observed number of individuals	f_1	f_2	f_3	... f_x
probability of capture	$\dfrac{m}{e^m}$	$\dfrac{m^2}{2e^m}$	$\dfrac{m^3}{6e^m}$	... $\dfrac{m^x}{x!e^m}$

The mean m is the probability of capture per trapping session, including those in which no animals were caught. The probability of no capture is $1/e^m$ (or e^{-m}, which is the same thing). This is unknown, because m is unknown. If we knew this value, however, we should know the population size, because it represents all the animals unseen during the sampling. The total number of animals caught once, twice etc; is f_1, f_2, etc. to f_s. Define the total number caught at least once as $r = \Sigma_{x=1}^{x=s} f_x$. The probabilities that they were caught once, twice, etc. sum to $1 - e^{-m}$. It follows that the ratio of P to r is $1 : 1 - e^{-m}$. Therefore, P can be found from:

$$\hat{P} = \frac{r}{1 - e^{-m}} \tag{5.12}$$

provided m can be estimated.

Once again, several methods are available. If the mean number of times an animal is captured is $\bar{x} (= \Sigma x f_x / r)$ then it may be shown (Craig 1953) that:

$$\bar{x} = \frac{m}{1 - e^{-m}} \tag{5.13}$$

This equation may be solved for m by trial and error; then substituting Equation 5.13 in Equation 5.12 we get for the estimate of P:

$$\hat{P} = \frac{r\bar{x}}{m}$$

$$= \frac{(\Sigma x f_x)^2}{m} \tag{5.14}$$

This method is applied to the *Clethrionomys* data in Note 5.4.

Note 5.4
Frequency of recapture method.

Example using the data for the Bank vole *Clethrionomys glareolus.*

The number of times each animal was captured was recorded as well as the other data discussed in Note 5.3. The results were as follows:

Times caught (x)	1	2	3	4	5	6	Total
Number caught (f_x)	21	13	12	1	3	1	$51 = \Sigma f_x$
Product ($x f_x$)	21	26	36	4	15	6	$108 = \Sigma x f_x$

Mean number of times captured:

$$\frac{\Sigma x f_x}{\Sigma f_x} = \frac{108}{51} = 2 \cdot 12 = \bar{x}$$

To test for randomness of distribution:

$$x^2 = \frac{\Sigma x^2 f_x - \dfrac{(\Sigma x_i f_x)^2}{\Sigma f_x}}{\bar{x}} = 37 \cdot 4$$

This has $\Sigma f_x - 1$ degrees of freedom. $P > 0 \cdot 05$, so that there is good agreement between the observed data and the Poisson expectation. This test has been found adequate in the truncated case involved here (David and Johnson 1952).

To find m:
A first approximation to m is:

$$m = \frac{\Sigma x^2 f_x}{\Sigma x f_x} - 1$$
$$= \frac{308}{108} - 1 = 1 \cdot 85$$

Let $y = \bar{x} - \dfrac{m}{1 - e^{-m}}$

Note 5.4 continued

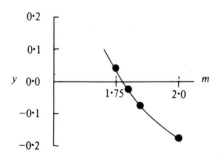

then by trial and error:

m	y
1·85	−0·075
2·00	−0·193
1·75	0·002
1·80	−0·036

Accepting $m = 1\cdot75$ as sufficiently accurate:

$$\hat{P} = \frac{\Sigma x f_x}{m} = \frac{108}{1\cdot75} = 61\cdot7$$

The data are derived from the same sample as those of Note 5.3. The difference in estimate arises because of the different assumptions being exploited by the two models.

Change in ratio of two classes after unequal removal

Animals of different types are sometimes removed from a population at different rates. The total sample removed is known. If the removal changes the ratio of the classes in a second sample of those which remain, then this fact may be used to estimate the population size. Here we have yet another approach to the manipulation of populations in order to estimate their numbers. When the classes distinguished are natural ones the method is most likely to be applied to populations subject to a management scheme, for example to populations of deer in which the males and the females are culled at different rates.

Suppose a population consists of a individuals of type A and b of type B, so that $a + b = P$. The observed frequency of type A in a sample is f_1, which is a measure of the fraction a/P. From the first sample, m of the A type and $n - m$ of the *B* type are removed, making a total of *n*. After this removal a second sample is observed, in which the frequency of type *A* is f_2. This is a measure of the fraction remaining after the removal. Thus:

$$f_2 = \frac{a - m}{P - n} \qquad (5.15)$$

But:

$$a = Pf_1 \qquad (5.16)$$

and rearranging these two equations we get:

$$\hat{P} = \frac{f_2 n - m}{f_2 - f_1} \qquad (5.17)$$

If all the individuals removed belong to the *A* class then $m = n$ and $\hat{P}$ becomes $n(1 - f_2)/(f_1 - f_2)$.

The method has been applied to deer and game bird populations where the two sexes are shot at different rates, to deer in which there is a known amount of differential mortality between young and adults, and to fish populations in which there is a lower limit to the size of individuals which may be taken.

Conclusions

This chapter introduces a variety of methods slightly different from the MRR, which illustrates the range of approaches available to the problem of estimating population size. They may be extended in a number of ways. Thus, trapping and removal methods may be modified to allow for variable catch effort. When *P* is estimated from increase in the fraction marked the estimate is distorted if the animals become more or less prone to recapture as time goes on (they become trap-happy or trap-shy). Methods are available which allow for this change. Similarly, when the method is based on frequency of recapture of individuals, a geometric (Eberhardt 1969) or a negative binomial (Holgate 1966, Caughley 1977a p. 153) rather than a Poisson distribution may be assumed. In each case the model becomes more realistic, but additional parameters may have to be estimated. The best up-to-date review of these options will be found in Seber (1973).

Sometimes data are available in such a form that only one method is applicable. The examples show, however, that different methods may be applied to the same data and yield different estimates of *P*. The *Clethrionomys* results indicate a population of 57 animals by the increase-in-fraction-marked method, but 62 by the frequency-of-recapture method. The difference is not great, but why does it arise and which figure is the best? In this case there is probably nothing to choose between them and the discrepancy is the result of sampling error. The assumption is that the probability of capture is constant with time and that captures are independent of each other. The slope of the curve of frequency of marks on cumulative samples is then linear and the frequency of capture follows a Poisson distribution. Accidental divergence has a slightly different effect on the two models.

The data on *Coenonympha pamphilus* have been analysed in two ways, yielding estimates of 544 and 510. The same remarks apply in this case. If two models are based on different assumptions then it is essential to understand the theory and relate it to the biology of the species concerned. To emphasise this point, consider a purely hypothetical example. Suppose birds are caught, banded and released as discussed on page 82. Ten birds are caught, of which 4 are recaptures, the sequence being 11111R1RRR. The maximum likelihood method derived from Equation 5.11 gives $P = 8.4$. If the regression method is used we get 8.2, which is not identical. In the maximum likelihood method only the number of birds (s) and the number of new captures (r) are used, whereas the regression method takes account also of the sequence in which recaptures are made (in the x_i column). All these variables may suffer accidental variation.

The standard error of the estimate is a measure of the probable variation to be expected for a given set of data. When other factors are equal, the best estimate to use is, therefore, the one which has the smallest standard error. The standard errors for the methods discussed are given in the next chapter.

6 Estimation of error

Introduction

In the previous chapters we have discussed examples in which alternative methods give different estimates when applied to the same data. This is because the methods start from different assumptions. All the assumptions are logical, but one set may be more appropriate to a particular set of data than another. Each method is based on a model that considers the probability of occurrence of certain events, so that random sampling also affects the estimate, and repeated samples from the same closed population using the same method would not give a succession of identical answers. The comparison of estimates obtained by different methods, therefore, involves two considerations: (a) the applicability of the model to the conditions prevailing and (b) the sampling error associated with the particular method. Since true population size is unknown, the esti-mated sampling error is usually the only indication of the confidence we can have in the estimate, and for this reason consideration of the errors of the esti-mates is important. These errors are discussed here, while the problem of choos-ing between different methods is reserved for Chapter 7.

The effect of sampling error

Suppose 50 individuals are caught and marked in a population of 100 individuals. They are returned and disperse randomly among the unmarked ones. A second sample of 10 is taken. For the Lincoln Index estimate we have $n_0 = 50, n_1 = 10$ and $P = n_0 n_1 / m_1$, where m_1 is the number of marked individuals in n_1. (This expression is the same as Equation 3.2, with $i = 0$.) Since P is 100 the average value of m_1 is expected to be 5, but the real number obtained may have any value from 0 to 10, governed by the probability of sampling 10 individuals at random from a population of 50 marked and 50 unmarked.

In this example the probability of picking a marked individual first time is 0.5. The probability of getting a marked individual then changes with each capture depending on whether the individual previously caught and removed is marked or unmarked, but in practice this complication may be ignored. The probability of getting m_1 marked individuals ($0 \leqslant m_1 \leqslant n_1$) can be taken to be:

$$\frac{n_1!}{m_1!(n_1-m_1)!}\left(\frac{1}{2}\right)^{m_1}\left(\frac{1}{2}\right)^{n_1-m_1} \tag{6.1}$$

For $n_1 = 10$ the calculated probabilities are:

m_1	0	1	2	3	4	5	6	7	8	9	10
probability	$\frac{1}{1024}$	$\frac{10}{1024}$	$\frac{45}{1024}$	$\frac{120}{1024}$	$\frac{210}{1024}$	$\frac{252}{1024}$	$\frac{210}{1024}$	$\frac{120}{1024}$	$\frac{45}{1024}$	$\frac{10}{1024}$	$\frac{1}{1024}$

Skewness of the estimate

In this example the probability distribution of m_1 in n_1 is symmetrical because the probability of getting a marked individual was 0.5; with other probabilities it would be skewed to the right or to the left. However, even here, the probable estimates of P are strongly skewed. Even though the most likely value of m is 5, for which $P = 100$, estimates of population size of 50 and of infinity are both equally likely ($m_1 = 10$ or 0 respectively). In Figure 6.1 results are given for 1000 samplings of a computer simulated population, under the conditions discussed. The frequency of estimates of each value of P reflect the probabilities of getting different numbers of marked individuals in n_1. The mode occurs at $P = 100$ but the skewed nature of the distribution is clearly seen. We shall return to this example in Chapter 7 (p. 101). For the moment the point to note is that the skewed distribution is characteristic of all population estimates. Even though a single standard error is given for an estimate, it does not imply that the probabilities of the estimate exceeding or being less than the true value are the same.

One approach to obtaining confidence intervals is to find the limits with respect to m_1 and then use these to calculate the equivalent values for P. This operation is similar to that discussed in Chapter 2 for static sampling, where the standard error of the mean number per quadrat was first calculated and then multiplied up to set the limits on the total. Let $a = m_1/n_1$. In a large sample the 95% confidence intervals for a are approximately $a \pm 2\sqrt{a(1-a)/n_1}$. Since $P = n_0/a$ the two values so obtained can be divided into n_0 to give equivalent 95% intervals for P. With $n_1 = 10$ and $m_1 = 5$ the confidence limits are $0.5 \pm 2\sqrt{0.025}$ or 0.19 to 0.81. The equivalent intervals for P are therefore 62 to 263. Even though the sample used is very small, comparison with the distribution in Figure 6.1 shows that the right kind of limits have been achieved. If confidence intervals are needed for MRR population estimates, a rule-of-thumb approach might be to proceed along these lines to work out the intervals for the estimated frequency of recapture based on the binomial or Poisson model, and to use them to calculate upper and lower limits for P.

Figure 6.1

Estimates of population size by the Lincoln Index for 1000 trials using a computer simulation. The true population size was 100, of which 50 were marked. Ten individuals were picked at random to produce each estimation. These results are also discussed in Chapter 7 and listed in Table 7.1.

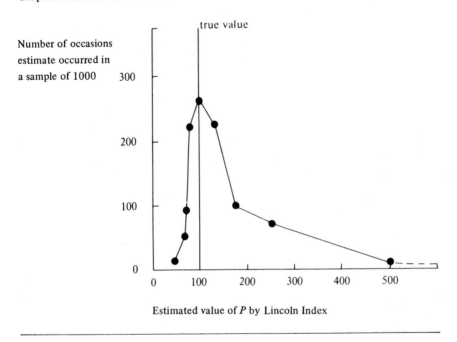

Estimated value of P by Lincoln Index

Sample and sub-sample size

For a Lincoln Index estimate n_1 will usually also be a random variable; for a constant effort put in, the sample taken may chance to be slightly larger or slightly smaller than expected. When discussing human population estimation by this method Laplace (1795) pointed out that n_1 must be at least 5% of P to get a reasonably accurate estimate of the population. However, since n_1 is not less than m_1, the effect of this variation is as small as or smaller than that of the variation in m_1. Such variation also affects the sampling error associated with P. When several different components are sampled to obtain the estimate the error is inflated, the component with the smallest sample size having the greatest effect. This point is discussed in Chapter 3 (p. 41) and Chapter 7 (p. 100).

A method for estimating errors

Just as the frequency of estimates of different size can be obtained in computer simulations using random sampling procedures, so the probability of errors of a given size for a single estimate may be calculated mathematically. The calculated parameter is the standard error, *SE*, of the estimate, which is the square root of its variance *V*. The standard error may be used to give an indication of the accuracy of an estimate, and is especially useful when comparing different sets of data, but as emphasised above it does not, as it stands, define confidence intervals above and below the estimated value. The method of estimation which is almost universally used in this field is called the method of maximum likelihood.

If we accept that the data available can yield the required estimate, (see Note 6.1), then the maximum likelihood method is the most efficient way of obtaining the estimate, and it also provides the estimated standard error. The method will now be outlined in relation to the Lincoln Index.

All individuals in the second sample of a Lincoln Index estimate must be either marked or unmarked. For unknown *P* the expression defining the probability of getting m_1 marks in n_1 is:

$$\frac{n_1!}{m_1!(n_1-m_1)!}\left(\frac{n_0}{P}\right)^{m_1}\left(1-\frac{n_0}{P}\right)^{n_1-m_1} \tag{6.2}$$

This expression cannot be evaluated to give the probability of some particular number of marks, because the probability n_0/P is unknown. We therefore refer to it as a likelihood function, as distinct from a probability function for which n_0/P or its equivalent would be given. In most circumstances very small or very large values of n_0/P lead to small probabilities, with larger probabilities lying between. A solution to the problem of estimating *P* is therefore to find the value of *P* for which the likelihood function is maximised, given n_0, n_1 and m_1.

The logarithm of this expression maximises at the same value of the probability n_0/P as the expression itself; this simplifies the calculation, since for any equation $y = a\ln x$ the derivative dy/dx is adx/x. In the present example the log likelihood is:

$$L = \ln\binom{n_i}{m_i} + m_1\ln\left(\frac{n_0}{P}\right) + (n_1-m_1)\ln\left(1-\frac{n_0}{P}\right) \tag{6.3}$$

and the maximum likelihood estimate of *P* is the value of *P* for which $dL/dP = 0$. Differentiating, we get:

$$\frac{dL}{dP} = -\frac{m_1}{P} + \frac{n_0(n_1-m_1)}{P(P-n_0)} = \frac{n_0 n_1 - m_1 P}{P(P-n_0)} \tag{6.4}$$

Note 6.1

The problem of whether the available data can yield the estimate may be exemplified by comparing two imaginary sets of binomial data. First, suppose 100 beads are taken from a bowl and found to consist of 70 red and 30 green beads. In the absence of any information on what was in the bowl a reasonable estimate of the probability of getting a red next time is 0·7. Secondly, suppose a coin is thrown one hundred times and falls heads on 70 occasions. By the same argument the probability of getting a head next time is 0·7. However, knowing the shape of the coin we may feel that the true probability is nearer to 0·5 and that the observed sequence is an unlikely one such as must only turn up very infrequently. The maximum likelihood method employs only the sample data, so is comparable to the first example.

Alternative methods of estimation also exist which include information, obtained before the event, such as the nature of the coin, as well as the sampling results. They have hardly ever been applied to population estimation problems, but are discussed, for example, by Gaskell and George (1972).

When $dL/dP = 0$ the numerator equals zero, so that $m_1 P = n_0 n_1$, or:

$$\hat{P} = \frac{n_0 n_1}{m_1} \qquad (6.5)$$

The usual Lincoln Index formula for P is therefore the maximum likelihood estimate, and represents the modal value of the probability distribution. The usefulness of this method is that it may further be shown that the variance of P is:

$$V_P = -\left(\frac{d^2 L}{dP^2}\right)^{-1} \qquad (6.6)$$

when this is evaluated with $P = n_0 n_1 / m_1$ (see Note 6.2). Differentiating expression (6.4) again and substituting for P we get:

$$V_P = \frac{P(P - n_0)}{m_1} = \frac{n_0^2 n_1 (n_1 - m_1)}{m_1^3}$$

The standard error of P is the square root of this expression.

In Table 6.1 variances for methods of estimation discussed in earlier chapters are listed.

Note 6.2

The argument outlined may be represented graphically in a form parallel to the algebra. The relation of L to P usually has a form something like the following:

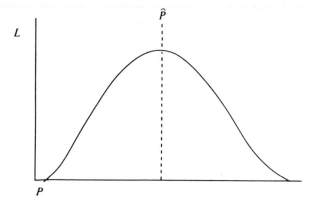

$\hat{P}$ is the value of P at the maximum value of L. The slope of L on P is dL/dP, which has the following form:

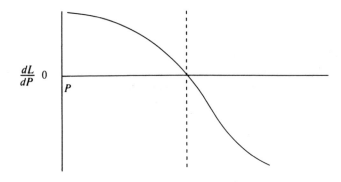

$\hat{P}$ is the value at which dL/dP is zero. The slope of dL/dP on P becomes steeper if the spread of L on P is reduced. It is therefore clearly related to the variance of P and, in fact, at $\hat{P}$ the slope is equal to $-1/V$. These relations indicate how $\hat{P}$ and $\hat{V}$ may be calculated numerically by iteration even when the algebraic equations have no explicit solution. If two pairs of values $x_1 y_1$ and $x_2 y_2$ are enumerated near $\hat{P}$ then $\hat{V} \cong (x_2 - x_1)/(y_1 - y_2)$.

Table 6.1
Variances for population estimates in Chapters 3, 4 and 5. In each case the standard error is the square root of the variance.

MRR Methods

Lincoln Index equation 3.2 page 29

$$V_{P_i} = \frac{n_i^2 n_{i+1}(n_{i+1} - m_{i+1})}{m_{i+1}^3}$$

Bailey's modification of Lincoln Index equation 7.1 page 101

$$V_{P_i} = \frac{n_i^2 (n_{i+1} + 1)(n_{i+1} - m_{i+1})}{(m_{i+1})^2(m_{i+1} + 2)}$$

Fisher and Ford Chapter 4, Note 4.3 page 60

Estimates of the variance of population size are not available. Leslie and Chitty (1951) showed, however, that the variance of the survival rate may be estimated from the rate of convergence of the values in *column 7* of Note 4.3 on zero as s moves towards $\hat{s}$. A numerical estimate of V_s, like that discussed in Note 6.2 may therefore be used.

For two values s_1 and s_2 lying on either side of s and close to it (e.g. $s - 0.01$ and $s + 0.01$) calculate the totals for *column 7*. Call these y_1 and y_2. Then:

$$V_s = \frac{s_1 s_2 (s_2 - s_1)}{s_2 y_1 - s_1 y_2}$$

Bailey's triple catch equation 3.23 page 40

$$V_{P_B} = P_B^2 \left(\frac{1}{a_1} + \frac{1}{b_1} + \frac{1}{a_2} + \frac{1}{B} \right)$$

Jolly equation 4.14 page 66

$$V_{P_i} = P_i(P_i - n_i) \left[\frac{M_i - m_i + n_i}{M_i} \left(\frac{1}{R_i} - \frac{1}{n_i} \right) + \frac{n_i - m_i}{n_i m_i} \right]$$

Table 6.1 continued

Jolly (1965) also gives an alternative equation for VP_i making more comprehensive assumptions. The difference is usually small. For the variances of the four estimations in Note 4.5 the average difference is a good deal less than 1%. The variance of the unbiased estimate (Ch. 7, p. 101) is the same as that above.

Manly and Parr   equation 4.17 page 69

$$VP_i = \frac{P_i Z_i (n_i - Y_i)}{Y_i^2}$$

Trapping and removal

Maximum likelihood method   equation 5.3 page 75

$$VP = \frac{Pq^S(1 - q^S)}{(1 - q^S)^2 - (1 - q)^2 s^2 q^{S-1}}$$

Regression method   equation 5.7 page 78

$$VP = \frac{\Sigma(n - \bar{n})^2}{p(s - 1)} \left[\frac{1}{s} + \frac{(P - \bar{x})^2}{\Sigma(x - \bar{x})^2} \right]$$

Increase in fraction marked

Regression method   equation 5.10 page 81

The error variance is

$$v = \left[\Sigma \frac{m_i^2}{n_i} - \frac{(\Sigma m_i x_i)^2}{\Sigma n_i x_i^2} \right] \frac{1}{s - 2}$$

The 95% confidence intervals are approximately:

$$\frac{\Sigma n_i x_i^2}{\Sigma m_i x_i \pm 2\sqrt{v \Sigma n_i x_i^2}}$$

Table 6.1 continued

Maximum likelihood method equation 5.11 page 82

$$V_p = \left[\frac{1}{P-r} + \frac{s-1}{P} - \sum_{i=1}^{i=s} \left(\frac{1}{P-n_i} \right) \right]^{-1}$$

Frequency of recapture equation 5.14 page 84

$$V_P = \frac{P}{e^m - 1 - m}$$

Confidence intervals could be obtained by working out the upper and lower 95% values of x for the Poisson distribution and substituting these for m in equation (5.13).

Change in ratio

$$V_P = \left[\frac{P^2 f_1(1-f_1)}{n} + \frac{(P-n)^2 f_2(1-f_2)}{n'} \right] / (f_1 - f_2)^2 \qquad \text{equation 5.17 page 87}$$

n' is the size of the second sample on which the observation f_2 is based.

Two end points

The methods of population estimation outlined are intended for situations where the investigator can hope to handle 10–50% of the population, which is likely to be numbered in the hundreds or perhaps thousands. Very large or very small populations pose their own problems.

Suppose a Lincoln Index estimation is made on a population and there are no recaptures. The answer we obtain is that the population is infinite. Obviously this is untrue, but is there any further information to be extracted from the sampling? Referring back to the probability function (6.2) we see that when $m_1 = 0$ the probability distribution becomes asymmetrical, maximising at $P = \infty$. In this case it would be reasonable not to maximise it but to calculate the value of P for which the expression has some arbitrary value. The value of P for which the probability is 0.5, for instance, is the one giving a 50 : 50 chance of obtaining no recaptures in n_1. With $m_1 = 0$ the expression then becomes:

$$\left(1 - \frac{n_0}{P}\right)^{n_1} = 0.5$$

so that:

$$\hat{P} = \frac{n_0}{1 - 2^{-(1/n_1)}}$$

If $n_0 = 50$ and $n_1 = 10$, as before, the population size which has a 50% chance of returning no recaptures is 747. Bell (1974, see also Edwards 1974) discusses the situation using the more accurate assumption that sampling progressively alters the size and composition of the remaining population.

At the other extreme, suppose the population is very small. Only one or two individuals are marked, they are returned to the habitat and every subsequent sighting turns out to be a marked individual. What length of sequence of sightings is needed to indicate with a given level of probability that the total population has, in fact, been marked? This problem was first discussed by Boguslavsky (1956) who provided a rule to determine the length of sequence consistent with a certain population at a given level of probability. Part of his sequence is reproduced below.

P	1	2	3	4	5	6	7	8	9	10
s	5	9	14	18	23	28	33	39	44	50

This shows the number of examinations (s) of the population required for 90% certainty that the population is no larger than P. Five successive sightings of a single animal are needed for 90% confidence that the population is not greater than one. If six different individuals have been marked then 28 sightings are needed to be confident that the population is not larger than six, and so on. Boguslavsky gives values up to $P = 15$, for which 79 sightings are required. With increase in P the method soon becomes tedious and the investigator is likely to resort to a different method of estimation (he may even invent one for himself!). Similar models derived by other authors are reviewed by Seber (1973).

Calculation

Apart from the work of Laplace (1783) and Petersen (1889), the literature on estimation begins in the 1930s, and there was a great upsurge of interest in the 1950s. Even ten years later than that the calculations were made using pencil and paper, with only slide rules, log tables and sometimes hand or simple electromechanical calculating machines as aids. As a result, many of the papers recommend forms of layout which reduce labour to a minimum (e.g. Fisher and Ford 1947) or approximations and mathematical expressions which are practical to calculate (e.g. Craig 1953). Several authors (e.g. Zippin 1956) provided nomograms allowing parameters to be read from a scale when the equations containing them have no explicit solution. Some of these are reproduced by Southwood (1966) and Seber (1973). We have not included them here, in the belief that most readers now have electronic calculators available which make them unnecessary.

Most of the operations outlined can be carried out on simple calculators, although the work is sometimes laborious. All are within the scope of the more versatile scientific machines. With these one can calculate $\ln x, e^x$ and non-integer roots and powers of x. A considerable advantage is gained if the machine is programmable, since parts of the operation such as the establishment of the Fisher and Ford survival rate may be programmed even if the complete calculation of P is not included. For the whole of a long calculation computer programs have much to recommend them. So long as the program has been thoroughly tested, its use minimises the chance of inaccuracies which otherwise tend to creep into repetitive calculations. For MRR estimations input of data in a contracted form of the Manly and Parr table allows all the other estimations discussed to be carried out as well, so that their comparative performance may be studied. Programming must be efficient to limit data storage to a minimum; even then a considerable store is required for a large data set.

At several points in this text we have included calculations which require iterative solution. These include the Fisher and Ford s (p. 58), Zippin's q (p. 75) and the m of the frequency of recapture method (p. 84). In each case the equations may be arranged in the form $y = f(x)$, where the value of $\hat{x}$ is the one for which $y = 0$. The relation of y to x is of the form seen in the second graph of Note 6.2. These equations may be solved simply by inserting trial values until a good agreement is reached, but a useful convergent method suitable for programming is outlined below.

Start with two values x_1 and x_2 not far from $\hat{x}$, and calculate y_1 and y_2. Then $x' = (x_2 y_1 - x_1 y_2)/(y_1 - y_2)$ is a better estimate of x. Substitute x' for x_1 and x_1 for x_2 and repeat the operation, stopping when $(x' - x_1)$ has diminished to some required small amount such as 0.001. This procedure results in rapid convergence, and in the case of the Fisher and Ford s, leaves one also with the values needed to estimate its standard error.

7 Choosing a method

Introduction

Several methods of population estimation are available: how do we decide which is the most suitable for a particular purpose? Sometimes animals are static, or can be treated as if they are by the use of suitable and rapid sampling methods such as aerial survey or photography. In such situations area samples can be used to estimate P as described in Chapter 2. This technique has the considerable advantage that a satisfactory estimate can be obtained when the sampling fraction (or sampling intensity) is much smaller than that required to give a reasonable value of P by use of MRR methods. This is advantageous if a large area or a dense population is under investigation. In some situations animals have to be killed when caught, and appropriate calculations that estimate P are discussed in Chapter 5. There are also a variety of approaches even when MRR is adopted. These differ from one another in several respects, which include (a) design for particular types of catching schedule, (b) assumptions about the pattern of survival, (c) the statistical approach underlying them, and (d) their relative efficiencies at different sampling intensities.

Decisions about (a) and (b) above depend on a knowledge of the animals concerned. Thus, if we believe the survival rate to be constant, and have to sample at irregular intervals for practical reasons, then the Fisher and Ford method is a good one to use. When the survival rate is known to be age-specific, Manly and Parr's method should be used if it is practical in other respects. With regard to the statistical approach the methods of Jolly, Seber and Manly & Parr are based on more realistic assumptions than the earlier methods, starting as they do by defining the probabilities with which certain events occur. However, since smaller categories are distinguished among the recaptures, they may be unusable at low sampling intensities when deterministic methods can still be employed (Ch. 6 p. 92). The investigator may have the choice of a poor estimate or no estimate at all. We therefore have to decide whether there are circumstances where the cruder methods actually perform better than theoretically more sophisticated ones.

It is usually impossible to determine relative effectiveness in the field since true population size and rates of gain and loss are unknown. A single comparison of a set of Fisher and Ford estimates with a known population has been made by Cook, Brower and Croze (1967), which indicated that the method gave satisfactory results. However, by simulating the estimation procedure on a computer we may measure the performance of different methods any number of times against populations with known size and other characteristics.

All techniques described in this book attempt to fit data to statistical models. They make the necessary but naïve assumption that individuals in a population are similar and have an equal probability of being captured during random sampling. This statistical ideal is rarely likely to be achieved exactly in nature and the importance of departures from it are difficult to assess. Two types of difficulty will arise. First there will be problems of sampling error that will be particularly acute when data are scarce. Secondly, there will be situations when data do not meet the requirements of a particular model. It is not easy to give general advice on these matters since the problems of sampling are as diverse as the populations on which these methods are intended to be used. The following paragraphs emphasise that one should be vigilant and thoughtful while collecting and analysing data. Table 7.3 is a synopsis of the properties and requirements of MRR models discussed in Chapters 3 and 4.

Inherent bias: a problem when data are scarce

Table 7.1 gives results from a very simple simulation of a closed population in which there is no birth or mortality and only one recapture occasion. The population size is 100, of which 50 individuals were marked. A second sample of 10 is selected at random from this population, and the sampling has been repeated one thousand times. In the 1000 trials the recaptures contained no marked individuals on one occasion, one on 8 occasions, two on 48, and so on. The mean number recaptured is 4.9, but the range is from 0 to 9, with a modal value of 5. The Lincoln Index is a method which estimates the mode of the probability distribution to which a large sample of values will tend. At the mode it provides $\hat{P} = 100$, a true reflection of the population size. However, if we calculate the arithmetic mean of the 999 estimates of P (excluding the first estimate of infinity) we find $\hat{P} = 116$, which is a slight overestimate. Theoretically the method is said to have a positive bias (Bailey 1951) of an estimatable amount. It causes most disturbance to the estimate when samples are small and contain few recaptures. As an alternative, when recaptures are few Bailey suggested that we calculate:

$$\hat{P}_i = \frac{n_i(n_{i+1} + 1)}{m_{i+1} + 1} \tag{7.1}$$

Bailey's correction counteracts the bias by the addition of 1 to the size of the second sample and to the number of recaptures in it. The equivalent expression for Jolly's method is:

$$P_i = \frac{(n_i + 1)\hat{M}_i}{m_i + 1}$$

Table 7.1

Estimates of population size for 1000 trials using a simple computer simulation. The true population size is 100 of which 50 are marked. Samples of 10 were taken ($SF = 0.1$). Using the numbers of marked recaptures obtained by random sampling, population estimates ($\hat{P}$) by the Lincoln Index, with and without Bailey's correction, have been calculated.

Number of marked recaptures	Number of trials	$\hat{P}$ Lincoln Index	$\hat{P}$ Bailey's correction
0	1	∞	550
1	8	500	275
2	48	250	183
3	112	167	138
4	219	125	110
5	267	100	92
6	220	83	79
7	84	71	69
8	35	63	61
9	6	56	55
10	0	–	–
4.9 Arithmetic mean		**116** (excluding first)	**101**

where

$$\hat{M}_i = \frac{(n_i + 1) Z_i}{R_i + 1} + m_i$$

(c.f. Table 3.6).

Using Bailey's corrected figure we get $\hat{P} = 92$ at the modal frequency but the mean of the repeated estimations is 101. By analogy, we may say of the single sample taken in the field that if Bailey's correction is used the estimate will be closer to the average of a hypothetical series of future samples than if the simple Lincoln Index is used. On the other hand, the single value which is most likely to occur is a slight underestimate. Bailey's correction is a useful one to apply when there are few recaptures, and may be included in multiple sample methods of analysis.

Moving on to more complex situations, Figure 7.1 gives estimates of the size (P) of simulated populations of 500 obtained by Jolly's and Fisher and Ford's methods (with and without Bailey's correction). Ten replicate populations were sampled on 10 occasions and a mean and standard error calculated for each time an estimate of P was possible. 20% of the population was taken on each sampling occasion and intersample survival rates were constant at 0.3, 0.4 or 0.5. The estimated survival rates are shown in Figure 7.2.

Estimates of the mean show that, as with the Lincoln Index, Bailey's correction appears to overcompensate for the bias particularly in Fisher and Ford estimates. However, the accompanying standard errors show that it improves consistency, particularly of the first one or two estimates of P which are more erratic and more biased than later ones. This occurs because the number of marks (m) in the population is increasing to an equilibrium determined by their rate of introduction and their rate of loss. In Figure 7.1 the balance is established within two days, but when there are higher survival rates it takes longer. These estimates alone appear to be improved by Bailey's correction, though in other circumstances where they are based on more recaptures (i.e. $m > 10$) even these estimates may not be seriously biased.

When few animals are recaptured (e.g. $m = 10$) because survival rate and sampling intensity are low (e.g. survival 0.3, sampling intensity 0.2) uncorrected Fisher and Ford estimates of P are more accurate and less prone to sampling error than those of Jolly. If m falls below 10, then corrected Fisher and Ford estimates may be best and a comparison of corrected and uncorrected figures will at least give some indication of the magnitude of the sampling errors. With better data (e.g. survival 0.5, sampling intensity 0.2, when m is approximately 20) there is little difference between the performance of the two models. Bailey's correction does increase the consistency of estimates, but this is sometimes hidden by the effect of chance in sampling too many or few recaptures. Gross disturbances will become less frequent either because of a higher sampling intensity or of a greater survival rate as m becomes greater than 20. When such complete data are available one has the luxury of choice of method of analysis (Table 7.3), though Jolly's model will usually provide satisfactory estimates as it calculates, and allows for, a separate survival rate during each intersampling period.

Table 7.2 summarises estimates of P obtained by MRR methods in Chapter 4. Although P is unknown some estimates are clearly less satisfactory than others because of deficiencies in the data and in the manner in which the calculations are made.

In Jackson's methods the adjustment of recapture values magnifies the sampling error associated with small numbers of captures. Entries for the hypotenuse cells only are used in the denominator to estimate P, so that some recaptures are ignored in estimating P. The large daily changes with the second method are due to the small number of entries used to calculate s.

Estimates of P obtained by the Lincoln Index (Note 4.2) are often based on less than 10 recaptures. Their positive bias is counteracted by Bailey's correction (figures in parentheses). The triple catch produces divergent values (e.g. successively 29 and 132) because of sampling error. The value of a_1 is 5 on day B; a value of 6 would have given $P_B = 76$. The Fisher and Ford estimate for day B is high because of sampling error, as is the Jolly estimate for day E. In the latter the last estimate in a series is particularly prone to error as R_E (Note 4.5) represents but one entry in the trellis. As has already been noted, Manly and Parr's method is sensitive to poor data because it distinguishes smaller categories among recaptures than other methods. This is particularly evident in estimates for days B and E. Note 4.6 shows that Y is only 1 and 2 respectively on these days. If an additional Y capture is assumed, values of P are markedly lower at 136 and 110.

The data were collected over 6 days under fairly consistent conditions (Fig. 4.1) and it is likely that the true survival rate did not fluctuate greatly from day to day. On balance, therefore, we would favour the Fisher and Ford estimates in this case. Survival rate may fluctuate, however, in which case the Jolly estimates are to be preferred.

Bias caused by failure of data to fulfil underlying assumptions of the model

Deficiencies under this heading are more serious in that a more restricted method of analysis may not correct them or reduce their effects.

Data for analysis by the Lincoln Index must fulfil several conditions if P is to be an accurate estimate. These conditions also apply to more advanced models except where specified allowances can be made. The assumptions made by, or

Table 7.2

A comparison of MRR estimates of P for the population of *Myrmeleotettix maculatus* discussed in Chapter 4. The Lincoln Index has been credited to the first day of each pair.

Day	Jackson First method s^-	Second method s^-	Lincoln Index		Triple catch	Fisher & Ford	Jolly	Manly & Parr
A			132	(115)				
B	102	69	72	(68)	91	107	81	231
C	55	110	81	(75)	69	61	79	72
D	62	29	98	(88)	29	63	73	84
E	76		122	(105)	132	77	102	147
F	94					82		

Figure 7.1

Means and standard errors of estimates of size obtained by Jolly's and Fisher & Ford's method for populations of 500 simulated by computer. Survival rates were as shown, sampling intensity was 0·2. Constant survival rates were used so that there was no undue bias against the Fisher and Ford model. Time-dependent changes in survival rate can be detected by homogeneity tests.

Standard errors are smallest when survival rate is 0·5. This indicates a greater consistency of the 10 estimates at 0·5 than at the lower rates of 0·4 and 0·3. Fisher and Ford values are more consistent at 0·3 than those of Jolly though the difference is much less marked at higher rates of survival. Bailey's correction reduces overestimates of P for the first 'day' of a series; it also improves consistency. (To convert standard errors to 95% confidence limits multiply by 2·23.)

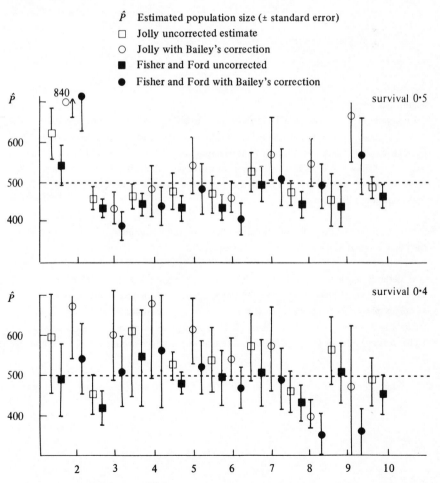

$\hat{P}$ Estimated population size (± standard error)
☐   Jolly uncorrected estimate
○   Jolly with Bailey's correction
■   Fisher and Ford uncorrected
●   Fisher and Ford with Bailey's correction

Figure 7.1 continued

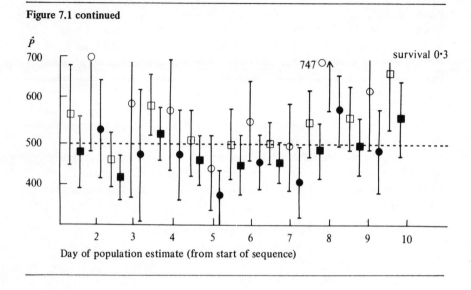

Day of population estimate (from start of sequence)

Figure 7.2

(a) A comparison between actual and estimated survival rates obtained by computer simulation. The single survival rate of the Fisher and Ford method and the arithmetic (A) and
geometric (G) means of estimates of 8 intersampling survival rates calculated by Jolly's
method are shown. (These latter estimates are with and without Bailey's correction.) Each
mean and standard error is based on 10 independent estimates. (Standard errors × 2·23 =
95% confidence limits.)

(b) Means and standard errors of estimates of survival rates obtained by Jolly's method for
the populations in Figure 7.1. (Standard errors × 2·23 = 95% confidence limits.)

The methods considered will resolve a difference in survival rate of 0·1 if the Fisher and
Ford survival or the mean values of Jolly's method are used. Individual estimates of intersampling survival from Jolly's method will not resolve a difference in survival of 0·1 and
frequently will not detect a difference of 0·2. The power of resolution will be greater at
higher survival rates or sampling intensities (e.g. if *m* becomes substantially greater than 20).

Figure 7.2 continued

s   Estimated survival rate (± standard error)

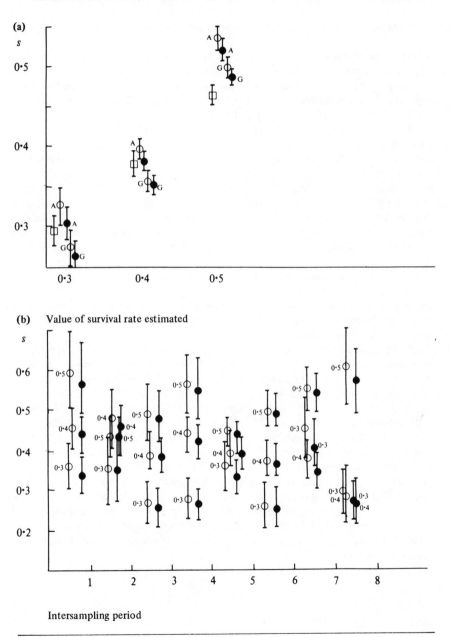

(a)

(b)   Value of survival rate estimated

Intersampling period

the conditions for, the applicability of recapture methods are noted at appropriate places in the previous chapters, especially Chapter 3 page 29. We now wish to return to them and to discuss methods of checking data to see whether the requirements are met. The conditions are as follows:

(i) Marking should not affect animals and marks should not be lost

The process of capturing and marking may be detrimental. If a unique mark is applied, it is possible to determine whether marking has an adverse effect since the 'recatchability' of animals caught for the first time at time x can be compared with recaptures released at the same time. This can be done by constructing a contingency table and calculating χ^2 (Note 7.1).

If marks are lost P will be overestimated. The likelihood and extent of mark loss can be checked by keeping a control population or by watching for indications of loss. An alternative method may be necessary or two methods may be used together.

Sometimes samples might be too large to mark by conventional individual or date specific marks and mass marking may be necessary. Occasionally the process of marking may be so laborious that few individuals can be handled. In these difficult circumstances it may be feasible to estimate some population parameters by Jackson's positive or negative method (see Table 7.3 at end).

(ii) The marked animals must mix randomly with the unmarked population

Sufficient movement has to occur to allow this to take place. The object is, all else being equal (see (iii)), that the population be resampled at random with respect to the mark status of its constituents.

If the power of movement of a species is unknown, a preliminary assessment is necessary. (For instance, of a sample released will some individuals move 5, 10, 50 or more metres in an intersampling period?) This will suggest how sampling effort should be distributed. A wide-ranging species will need a broad geographical spread of effort to obtain a reasonable return of marks. A species with limited powers of dispersion will require locally distributed sampling. The moths *Biston betularia*, which can fly several kilometres in a night, and *Gonodontis bidentata*, which flies about 100–200 metres in a similar period, are examples at opposite ends of this scale. Where movement is limited and sampling is random, release of marked animals at the place of their capture would help ensure that marked and unmarked animals were taken at random at the next sampling occasion. It might also be possible to increase the intersampling period to allow marks longer to redistribute. A specified release area should only be used if animals move actively enough to redistribute themselves amongst the whole population.

(iii) Samples must be obtained at random: an animal's age, sex or mark (also see (i) and (ii)) should not influence its chance of capture

This condition places severe limitations on the data that are suitable for analysis. If all individuals are not equally catchable, vast errors can occur. If previously marked individuals avoid recapture, then the population will be overestimated; if they are prone to recapture (e.g. become addicted to traps), the population will be underestimated. Tests of the randomness of sampling can be applied (see examples of Leslie's test, Note 7.2) but they are insensitive and the absence of a significant result does not always imply true randomness of sampling.

Obviously data for analysis should be taken from populations that are as homogeneous as possible in order to reduce variation in the probability of capture. This may demand separate analysis of subpopulations of each sex, instar or genetically determined morph. The division of data into subpopulations may reduce the number of recaptures on an occasion and alter one's choice of model.

If the probability of death increases with age, then it may be possible to use Manly and Parr's model which makes allowance for such loss. This method is, however, unreliable when the number of animals known to be alive but not captured is small. In natural populations loss is frequently age-independent; however, a subjective way of checking data is to construct a composite life table. If loss is age-independent the number surviving should decline exponentially with age, i.e. the log of frequency in class against time should produce a straight line (see example, Note 7.3).

(iv) Sampling must be performed at discrete intervals and the time spent sampling must be small in relation to the intersampling periods

Animals must be allowed time to undertake their usual activities and be exposed to the normal risks of life. Processes which interfere with this (e.g. overnight trapping of rats and mice) may severely restrict both and have a complex, though not readily analysable, effect on estimates. In the case of rodents this problem is alleviated by extending the intersampling periods, which has the desirable side effect of reducing the influence of trap addiction and trap avoidance. This solution will not be possible with those species that have a short expectation of life; in such cases it might be appropriate to restrict the sampling period and use a more restricted method to be determined by the quality of data (Table 7.3).

Note 7.1

Identifying homogeneous groups within capture–recapture data.

If each animal in a sample is given a unique mark at first capture it can be determined whether the process initially increases chances of death. A 2 × 2 contingency table and χ^2 test shows whether such loss occurs. Male rats were captured in hedges in farmland (Bishop & Hartley 1976) and were classified in the categories shown in the table. The homogeneity of the two groups in the left-hand column with respect to their fate after release (top row) was examined by means of a χ^2 test. The comparison is between rats captured and released for the first time (when they were uniquely marked by clipping off two toes) and those being released after first recapture (when rats were treated in the same way as at first capture except they were not marked).

	Number not recaptured	Number recaptured
Rate after first capture	92	43
Rate after first recapture	19	12

$$\chi_1^2 = 0.535$$

The value of χ^2 would be expected by chance and we can conclude that marking did not increase the loss after first capture.

Such 2 × 2 tables can test the homogeneity of many selected groups of data (e.g. relating to recatchability of sexes, age classes, morphs or of otherwise similar individuals released on different occasions). A large series of tests is more sensitive to biological change than a single larger one, and individual comparisons can be related to points in time within a sequence of population estimates. White (1975) shows the power of such an approach in dealing with capture–recapture data from populations of grasshoppers.

Note 7.2

Test of randomness of sampling.

Leslie's test (Leslie 1958) examines the frequencies of recapture of individuals. If 'catchability' is constant (i.e. if sampling is random) such frequencies will conform to a binomial distribution. This test requires data from 20 or more animals known to have survived at least four sampling occasions.

Note 7.2 continued

Leslie's original data was obtained from a population of sea birds (shearwaters) sampled
once each year. Thirty-two birds marked in 1946 were recaptured in 1952. The population
was sampled five times in the interim so that each bird marked in 1946 could have been
captured a maximum of five times between 1946 and 1952.

Year of sample	Recaptures (n)	Times recaptured (i)	Number of animals (f)
1947	7	0	15
1948	7	1	7
1949	6	2	7
1950	4	3	2
1951	7	4	1
		5	0
	$\Sigma n = 31$		$\Sigma f = 32$
	$\Sigma n^2 = 199$	$\Sigma fi = 31$	$\Sigma fi^2 = 69$

The column of recaptures (n) shows the number of these handled each year, irrespective
of their past histories. The two other columns show the number not caught during the
period, caught once, twice etc. up to five times. The observed variance of recaptures can
then be compared with the expected binomial using the expression:

$$\chi^2 = \frac{\Sigma fi^2 - \dfrac{(\Sigma fi)^2}{\Sigma f}}{\dfrac{\Sigma fi}{\Sigma f} - \dfrac{\Sigma n^2}{(\Sigma f)^2}} = 50 \cdot 32$$

degrees of freedom $(\Sigma f) - 1 = 31$.

This is significant $(P < 0\cdot02)$, so the data suggest that shearwaters are not being sampled at
random.

The test has the disadvantage that it is unable to use information from the first and last
sampling occasions and from animals not caught on the last occasion. This is because it
requires a defined group that is known to be alive throughout the period under investigation.
Although data for different sampling periods can be combined, this restriction seriously
limits the data that can be used, which will frequently be insufficient for the test. If the
interval between the first and last sample is short enough to ensure that loss (death plus
emigration) is negligible, all capture records can be used and data compared with a zero-
truncated Poisson distribution (see Note 5.4). If sampling is random there should be no

Note 7.2 continued

appreciable differences between the observed data and the expectation predicted by the distribution. The test also requires that the number of sampling occasions be large in relation to the mean number of times an animal is recaptured. This constraint, together with the assumption of negligible loss, will restrict the usefulness of the technique to some vertebrate populations (see Caughley 1977a, pp. 137–8, for method).

Note 7.3
Age-dependent mortality.

If animals can be aged, a full life table can be constructed. If necessary, the number of animals recaptured from a particular group can be compared with similar data from another group (see p. 110). Significant heterogeneity will indicate that loss is age dependent. The problem will then be to decide whether the data are sufficiently complete to be analysed by Manly and Parr's method (Table 7.3).

When it is not possible to age an animal directly an approximate indication can be obtained from the length of known life in a capture–recapture experiment. Young animals are more likely to be marked for the first time than older ones.

A composite life table was constructed from data obtained in a capture–recapture experiment which investigated a scalloped hazel moth (*Gonodontis bidentata*) population in England (Bishop, Cook & Muggleton 1978):

Length of known life (days)	1	2	3	4	5	6	7	8	9	10
No. in 'age' category	125	52	25	9	5	5	2	0	2	1
Natural logarithm of no.	4·82	3·95	3·21	2·10	1·60	1·60	0·69	–	0·69	0·00

If loss is age independent, the numbers should decline in a negative exponential manner and their logarithms should fall along a straight line. The graph shows that this is so, though there are some irregularities after day 6, probably because of sampling error. Similar data for the butterfly, *Heliconius charitonius*, in Costa Rica (Cook, Thomason & Young 1976) are also presented. The change in slope after age 70 days indicates age-dependent loss.

Note 7.3 continued

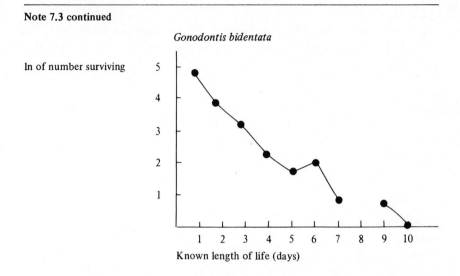

Gonodontis bidentata

ln of number surviving

Known length of life (days)

If a less subjective check of age independence is required, a χ^2 or exact test of homogeneity of loss from two or more 'age' classes can be performed as suggested above.

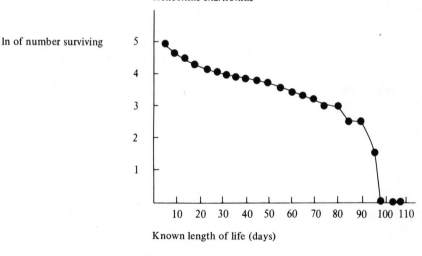

Heliconius charitonius

ln of number surviving

Known length of life (days)

Table 7.3
Summary of properties of MRR models discussed in the text. It is essential that animals involved in multiple recapture studies should be given a date-specific mark at every capture or an individual mark.

Model	Number of samples required	Rate of loss	Rate of gain
Lincoln Index	2	0	0
		constant	0
		0	constant
Bailey's triple catch	3	constant	constant
Jackson	3 or more		
+ ve		variable	constant
— ve		constant	variable
Fisher & Ford	3 or more	constant	variable

Table 7.3 continued

Parameters estimated	Remarks
$P_1 = P_2$ P_1 P_2	to be used when only two samples are available; use Bailey's correction when $m < 10$
(a) P_2 (b) survival rate (c) rate of gain	the minimum number of captures required to allow for constant loss and gain: it is instructive to compare the theories of the different methods in the triple catch, but best to use Jolly's in practice
(a) P for every sample except last (b) single constant gain rate (c) loss between each estimate of P	records change in proportions of marked animals in successive recaptures of one day's mark; (a restricted method useful when a large number of animals can easily be mass marked, e.g. *Drosophila* marked with fluorescent dust); adjustment of recapture values magnifies the error when captures are few
(a) P for every sample except first (b) single constant survival rate (c) gain between each estimate of P	records relative proportions of marks applied on successive days and recaptured on one day; (a restricted method useful when there is a constraint on numbers that can be marked at one time)
(a) P for every sample time except first (b) single constant survival rate (c) gains between each estimate of P	model calculates a single survival rate which is assumed to remain constant for duration of experiment; with high survival rates place less reliance on first estimates of P in a series than later ones; useful at relatively low sampling intensities, or with low survival rates when $m < 10$; calculate with or without Bailey's correction

Table 7.3 continued

Model	Number of samples required	Rate of loss	Rate of gain
Jolly	3 or more	variable	variable
Manly & Parr	3 or more	variable	variable

Fisher and Ford's method uses data from recaptures relatively inefficiently. A more efficient method which assumes a constant survival rate is desirable. Such a method has been developed by G. M. Jolly (*Biometrics* in press).

A computer program is available from Manchester University which carries out all the above calculations starting from the Manly and Parr trellis.

Table 7.3 continued

Parameters estimated	Remarks
(a) P for every sample time except first and last (b) survival rate for each intersample period except last (c) gains between each estimate of P	calculates separate survival rates for intersampling periods; these assume that every animal has the same chance of surviving until next sample (i.e. mortality is independent of animals' age); last estimate of P is prone to sampling error; use when $m > 10$, preferably when $m > 20$, usually the best method when s can not be assumed to be constant
as by Jolly's method	calculates a separate survival rate for intersampling periods; these do not assume that mortality is independent of age; particularly sensitive to sampling error because of the small categories distinguished (especially the quantity Y); smallest category should be 10 or greater, which in practice requires high sampling intensity and survival rate

The closed population

If animals can be marked and the population is a closed one, then the Lincoln Index or one of the methods of Chapter 5 (other than trapping and removal) should be used. Gain and loss must be negligible but the requirements of the previous section must be met. When the population is open we have seen that there are many obstacles to adequate MRR analysis which can be negotiated with careful treatment of the data. In some instances practical experience tells us that the data will still be unsatisfactory, or that the amount of effort required will be unreasonably large. In that case it may be possible to treat the population as effectively closed during part of the sampling period. An example is given in the introduction to Chapter 5, where a trapping and removal method is substituted for an MRR method. Gain and loss would have been substantial over the intersampling period required for the MRR study, but are negligible over a period sufficient for the trapping and removal estimate.

List of symbols used

Our first aim has been consistency within the book. In the MRR sections we have started from basic principles using a simple notation and developed the argument and a more general notation together. The symbols differ slightly from those of Seber (1973), who has used a standardised notation.

Symbol		Chapter
A	Sampling area, total area to which a sample is ascribed	(Ch. 2, p. 18)
a	Total area of sample	(Ch. 2, p. 18)
A, B, C	Number of animals captured, marked and released on the first three days of a MRR programme	Values for triple catch of the
a, b, c	Recaptures of marks applied	general n_i and m_i
a, b, c	Recaptures of marked animals carrying single marks	(Ch. 3, p. 34
ab	Recaptures of animals marked on both day i and day $i + 1$	et seq.)
b	Slope of regression of y on x	(Ch. 5, p. 80)
C	Constant in a probability expression	(Ch. 5, p. 75)
d	Death rate of animals dying *in situ*; the usual parameter measured in the rate of loss $(= 1 - s)$, which is composed of death plus emigration	(Ch. 4, p. 71)
e	Base of natural logarithms	(Ch. 2, p. 22)
e	Emigration rate (see above)	(Ch. 4, p. 71)
	Expectation of life, e_0 at birth, e_x at age x	(Ch. 4, p. 72)
f	Frequency; f_x, the number caught x times	(Ch. 5, p. 84)
	the number of sampling units containing x animals	(Ch. 2, p. 22)
	the fraction of a given type in a sample	(Ch. 5, p. 86)
g	Gain to a population (ingress, or births plus immigration) as a number for a specified interval	(Notes 4.2, 4.5 and 4.6)
i, j	Subscripts to define particular values in a series of values	
k	Total number of values in a sequence	(Ch. 5, p. 78)
L	Log likelihood	(Ch. 6, p. 92)

M	Number of marked animals (Jolly, Manly & Parr) or marks (Jackson, Fisher & Ford) in a population M is usually an estimate $\hat{M}$	(Ch. 3, p. 39)
m	Number of marked animals (Jolly, Manly & Parr) or marks (Jackson, Fisher & Ford) in a sample, i.e. the number of recaptures or number marked previously	(Ch. 3, p. 29) (Ch. 5, p. 80)
	Mean number per sample unit	(Ch. 2, p. 22)
	Number in one class in change in ratio method	(Ch. 5, p. 86)
N	Number of sample units (quadrats) in a sample	(Ch. 2, p. 22)
n	Number of animals in a sample unit or sample	(Ch. 2, p. 19; Ch. 3, p. 29; Ch. 5, p. 75)
P	Population size, usually the estimate $\hat{P}$	(all chapters)
p	Probability of capture or of appearing in a sample unit	(Ch. 5, p. 75)
q	Probability of not being caught $(= 1 - p)$	(Ch. 5, p. 75)
R	Number of n_i animals subsequently recaptured	(Ch. 4, p. 64)
r	Number of animals captured in two successive samples in Manly and Parr method (r_i is the number caught on i and $i + 1$)	(Ch. 4, p. 69, Note 4.6)
	Total number of different animals caught	(Ch. 5, p. 82)
	Number caught at least once	(Ch. 5, p. 84)
s	Survival rate for a given unit of time, usually from i to $i + 1$. In Jackson's method s^- (s elsewhere) is the conventional survival rate, s^+ is the rate resulting from dilution of the population by gain	(Ch. 3, p. 35 and elsewhere)
	Number of successive samples	(Ch. 5, p. 75)
SE	Standard error	(Ch. 2, p. 23; Ch. 6, p. 92)
SF	Sampling fraction, usually equivalent to the probability of capture	(Ch. 2, p. 18; Ch. 3, p. 30)
SI	Sampling intensity $(= SF)$	
T	Total number of units into which a sample area is divided	(Ch. 2, p. 26)
	Total days survived by all marks in Fisher and Ford	(Ch. 4, p. 60, Note 4.3)
V	Variance	(Ch. 2, p. 23; Ch. 6, p. 92)
X	Standardised number of captures in Jackson's method	(Ch. 4, p. 52)
X	Σx, where x_i is an animal having its first or last appearance in a sample on day i	(Ch. 3, p. 37; Ch. 4, p. 45, Note 4.1)
x	Number of animals in a quadrat	(Ch. 2, p. 22)
	total animals unmarked	(Ch. 5, p. 80)
$\bar{x}$	Mean number of times an animal is caught; similarly, $\bar{y}$ (Ch. 5, p. 80) and $\bar{n}$ (Note 5.2, p. 79)	(Ch. 5, p. 84)

are means of the series of numbers y_i and n_i

Y Σy, where y_i is an animal captured on day i and also before and subsequently   (Ch. 3, p. 37; Ch. 4, p. 45, Note 4.1)

y Cumulative fraction of animals marked   (Ch. 5, p. 80)

Z Σz, where z_i is an animal uncaught on day i but caught before and subsequently   (Ch. 3, p. 42; Ch. 4, p. 45, Note 4.1)

z Area of a sample unit, when this is variable   (Ch. 2, p. 26)

Glossary

birth	See **gain**
captures	Individuals caught and comprising a **sample**. In **MRR methods** we have assumed for simplicity that all captures are released. This is not necessary, and the equations may be adjusted to allow for individuals killed on capture or retained for other purposes.
day	Used for the time interval between successive captures. Any appropriate time unit.
death	See **loss**
deterministic method	A method of estimation in which the initial assumptions do not include random fluctuation. For **MRR**, the methods of Jackson and Fisher & Ford.
dilution	Reduction in the marked fraction of a **population** by **gain**.
egress	**Loss**
emigration	See **loss**
expectation of life	Mean number of time units survived. If there is constant survival rate s per unit time, the expectation of life at the start is $-1/\ln s$.
gain	Input to a **population** as a result of **immigration**, **birth** or emergence from a previous life stage.
immigration	See **gain**
ingress	**Gain**
loss	Reduction of a **population** as a result of **death** or **emigration**. Usually expressed as a rate per unit time, when the complement is survival rate.
mark	Any record used for recognising an individual. Marks may be individual, such as a unique applied number or a natural character (finger print), or they may be date specific, such as coloured paint or dye spots. If the latter, an individual may carry several marks. Individual marking sometimes

MRR method

reduces handling time, but all the information required is provided by date-specific marks.
A method of estimation by marking, release and recapture of individuals. The required population estimate is obtained by equating a ratio containing the unknown P with another ratio which can be estimated from the data.

maximum likelihood method

Statistical estimation method in which a probability function is defined and then maximised, given the observed **sample** data.

population

The number of individuals using a defined area during the period of investigation. Individuals may spend their whole lives in the area (e.g. the population of tortoises on a small oceanic island) or use may be temporary (e.g. the population of antelopes using a water hole to drink). A population is closed if it remains unchanged by **loss** or **gain** during the investigation, open if **loss** and/or **gain** occur.

releases

See **captures**

sample

Part of a **population** which is captured and from which parameters of the uncaptured fraction are estimated.

stochastic method

A method of estimation in which the probabilities of events occurring are defined in the initial assumptions. For **MRR**, the methods of Jolly and Manly and Parr.

trellis

A triangular matrix of numbers of captures and recaptures defined by co-ordinates of date of release and capture. Type 1 trellis includes all **marks** recaptured, whether carried singly or in combination; type 2 contains marked individuals recaptured, usually entered for the date of the most recently applied **mark**.

References

Ashford, J R, K L Q Read and **G C Vickers** 1970. A system of stochastic models applicable to studies in animal population dynamics. *J. Anim. Ecol.* **39**, 29–50.

Bailey, N T J 1951. On estimating the size of mobile populations from recapture data. *Biometrika* **38**, 293–306.

Bell, G 1974. Population estimates from recapture studies in which no recaptures have been made. *Nature* **248**, 616.

Bishop, J A, L M Cook and **J Muggleton** 1978. The response of two species of moths to industrialization in north-west England. II Relative fitness of morphs and population size. *Phil. Trans Roy. Soc. Lond. B* **281**, 517–42.

Bishop, J A and **D J Hartley** 1976. The size and age structure of rural populations of *Rattus norvegicus* containing individuals resistant to the anticoagulant poison Warfarin. *J. Anim. Ecol.* **45**, 632–46.

Bishop, J A and **P M Sheppard** 1973. An evaluation of two capture-recapture models using the technique of computer simulation. In *The Mathematical theory of the dynamics of biological populations*, M S Bartlett and R W Hiorns (eds), 235–52. New York and London: Academic Press.

Boguslavsky, G W 1956. Statistical estimation of the size of a small population. *Science* **124**, 317–18.

Caughley, G 1977a. *Analysis of vertebrate populations*. New York: John Wiley.

Caughley, G 1977b. Sampling in aerial survey. *J. Wildl. Manag.* **41**, 605–15.

Chapman, D G 1954. The estimation of biological populations. *Ann. Math. Statist.* **25**, 1–15.

Chapman, D G and **G I Murphy** 1965. Estimates of mortality and population from survey-removal methods. *Biometrics* **21**, 921–35.

Cochran, W G 1977. *Sampling techniques,* 3rd edn. New York: John Wiley.

Cook, L M, L P Brower and **H J Croze** 1967. The accuracy of a population estimation from multiple recapture data. *J. Anim. Ecol.* **36**, 57–60.

Cook, L M, E Thomason and **A M Young** 1976. Population structure, dynamics and dispersal of the tropical butterfly *Heliconius charitonius*. *J. Anim. Ecol.* **45**, 851–63.

Cormack, R M 1969. The statistics of capture–recapture methods. *Oceanogr. Mar. Biol. Ann. Rev.* **6**, 455–506.

Cormack, R M 1973. Commonsense estimates from capture—recapture studies. In *The Mathematical theory of the dynamics of biological populations*, M S Bartlett and R W Hiorns (eds), 225–34. New York and London: Academic Press.
Craig, C C 1953. On the utilization of marked specimens in estimating populations of insects. *Biometrika* **40**, 170–6.

Darroch, J N 1961. The two-sample capture—recapture census when tagging and sampling are stratified. *Biometrika* **47**, 241–60.
David, F N and **N L Johnson** 1952. The truncated Poisson. *Biometrics* **8**, 275–85.
Davies, R G 1971. *Computer programming in quantitative biology*. New York and London: Academic Press.
Debauche, H R 1962. The structural analysis of animal communities of the soil. In *Progress in soil zoology*, P W Murphy (ed.). London: Butterworth.
De Lury, D B 1951. On the planning of experiments for the estimation of fish populations. *J. Fish. Res. Bd Canada* **8**, 281–307.
Dowdeswell, W H, R A Fisher and **E B Ford** 1940. The quantitative study of populations in the Lepidoptera. I *Polyommatus icarus* (Rott.). *Ann. Eugen.* **10**, 123–36.

Eberhardt, L L 1969. Population estimates from recapture frequencies. *J. Wildl. Manag.* **33**, 28–39.
Edwards, A W F 1974. Population estimates from recapture studies. *Nature* **252**, 809–10.
Elliott, J M 1971. *Some methods for the statistical analysis of samples of benthic invertebrates.* Freshwater Biological Association Scientific Publication No. 25. The Ferry House, Ambleside, Cumbria, U.K.

Fisher, R A and **E B Ford** 1947. The spread of a gene in natural conditions in a colony of the moth *Panaxia dominula* L. *Heredity* **1**, 143–74.

Gaskell, T J and **B J George** 1972. A Baysian modification of the Lincoln Index. *J. appl. Ecol.* **9**, 377–84.
Grieg-Smith, P 1964. *Quantitative plant ecology*. London: Butterworth.

Hammersley, J M 1953. Capture—recapture analysis. *Biometrika* **40**, 265–78.
Hayne, D W 1949. Two methods for estimating populations from trapping records. *J. Mammal.* **30**, 399–411.
Hilborn, R, J A Redfield and **C J Krebs** 1976. On the reliability of enumeration for mark and recapture census of voles. *Can. J. Zool.,* **54**, 1019–24.
Holgate, P 1966. Contributions to the mathematics of animal trapping. *Biometrics,* **22**, 925–36.

Jackson, C H N 1933. On the true density of tsetse flies. *J. Anim. Ecol.* **2**, 204–9.

Jackson, C H N 1939. The analysis of an animal population. *J. Anim. Ecol.* **8**, 238–46.

Jackson, C H N 1948. The analysis of a tsetse fly population III. *Ann. Eugen.* **14**, 91–108.

Jolly, G M 1965. Explicit estimates from capture–recapture data with both death and immigration – stochastic model. *Biometrika* **52**, 225–47.

Jolly, G M 1969. Sampling methods for aerial censuses of wildlife populations. *E. Afr. Agric. For. J.* **34**, 46–9.

Laplace, P S 1783. Sur les naissances, les mariages et les morts à Paris, depuis 1771 jusqu'en 1784, et dans tout l'étendue de la France, pendant les années 1781 et 1782. *Mem. Acad. royale des Sciences de Paris* in Vol. 11: *Oeuvres complètes de Laplace*. Paris 1912: Gauthier-Villars.

Laplace, P S 1795. in Vol. 14: *Oeuvres complètes de Laplace* p. 164. Paris 1912: Gauthier-Villars.

Le Cren, E D 1965. A note on the history of mark–recapture population estimates. *J. Anim. Ecol.* **34**, 453–4.

Leslie, P H 1958. Statistical appendix. *J. Anim. Ecol.* **27**, 84–6.

Leslie, P H and **D Chitty** 1951. The estimation of population parameters from data obtained by means of the capture–recapture method. I. The maximum likelihood equations for estimating the death rate. *Biometrika* **38**, 269–92.

Leslie, P H and **D H S Davis** 1939. An attempt to determine the absolute number of rats in a given area. *J. Anim. Ecol.* **8**, 94–113.

Lincoln, F C 1930. *Calculating waterfowl abundance on the basis of banding returns.* US Dept Agric. Circ. No. 118, 1–4.

Manly, B F J and **M J Parr** 1968. A new method of estimating population size, survivorship, and birth rate from capture–recapture data. *Trans Soc. Brit. Ent.* **18**, 81–9.

Moran, P A P 1951. A mathematical theory of animal trapping. *Biometrika* **38**, 307–11.

Norton-Griffiths, N 1978. *Counting animals.* Handbook No. 1, 2nd edn. Nairobi. Serengeti Ecological Monitoring Programme of African Wildlife Leadership Foundation.

Parr, M J 1965. A population study of a colony of imaginal *Ischneura elegans* (van der Linden) (Odonata: Coenagriidae) at Dale, Pembrokeshire. *Fld Studies* **2**, 237–82.

Paulik, G J and **D S Robson** 1969. Statistical calculations for change-in-ratio estimations of population parameters. *J. Wildl. Manag.* **33**, 1–27.

Petersen, C G J 1889. *Fisk. Beretn. Kbh.* 1888–9.

Pielou, E C 1977 *Mathematical ecology.* New York: Wiley.

Roff, D A 1973. An examination of some statistical tests used in the analysis of mark–recapture data. *Oecologia* **12**, 35–54.

Schnabel, Z E 1938. The estimation of the total fish population of a lake. *Amer. Math. Mon.* **45**, 348–52.

Schumacher, F X and **R W Eschmeyer** 1943. The estimation of fish populations in lakes and ponds. *J. Tennessee Acad. Sci.* **18**, 228–49.

Seber, G A F 1965. A note on the multiple–recapture census. *Biometrika* **52**, 249–59.

Seber, G A F 1973. *The estimation of animal abundance.* London: Griffin.

Southwood, T R E 1966. *Ecological methods with particular reference to the study of insect populations.* London: Methuen. Second edn., 1978. London: Chapman & Hall.

White, E G 1975. Identifying population units that comply with capture–recapture assumptions in an open community of alpine grasshoppers. *Res. Popul. Ecol.* **16**, 153–87.

Zippin, C 1956. An evaluation of the removal method of estimating animal populations. *Biometrics*, **12**, 163–9.

Index